The Non-Nuclear Defense of Cities:

The High Frontier Space-Based Defense Against ICBM Attack

General Daniel O. Graham

Abt Books
Cambridge, Massachusetts

Library of Congress Cataloging in Publication Data

Graham, Daniel Orrin, 1925–
The non-nuclear defense of cities.

1. Antimissile missiles. 2. United States—Air defenses, Military. 3. United States—Military policy. I. Title. II. Title: High frontier.
UG1312.A6G73 1983 358'.8'0973 83-2534
ISBN 0-89011-586-9

Printed in the United States of America.

TABLE OF CONTENTS

FIGURES

FOREWORD

Lt. Gen. Daniel O. Graham USA (Ret.)

High Frontier is a privately funded effort, conducted under the aegis of The Heritage Foundation. Its purpose is to seek answers in U.S. technology, especially space technology, to the strategic problems that plague the United States and the Free World.

The origins of the effort lie back in the days when I was a military advisor to then-candidate Ronald Reagan. Early in the campaign I was among those insisting that the only viable approach for a new administration to cope with growing military imbalances was to implement a basic change in U.S. grand strategy and make a "technological end-run on the Soviets."

As far as I could determine, all advisors to Mr. Reagan agreed with this conclusion at least in principle at the time. However, as time passed, this more fundamental approach to national security issues receded into the realm of theory; the team of advisors on security matters began to concentrate instead on the amounts of money needed to revitalize ailing, ongoing Pentagon programs and on the ''quick fixes'' necessary if the United States were to hold its own within the context of *current* strategy and doctrine. The Carter defense budget was gone over line item by line item with a view to repairing past damage to U.S. capabilities with increased resources. New program expenditures were recommended to plug as quickly as possible the strategic gaps between U.S. and Soviet capabilities which are known collectively as ''the window of vulnerability.''

Some of the team continued to believe that a strategically and technically sound alternative to this incremental approach could be found, but none of us were quite sure at the time what the alternative might be.

In early 1981, Congressman Newt Gingrich of Georgia and I discussed the future of the new Administration in the national security field. Mr. Gingrich shared my apprehension that large Department of Defense budget increases alone would not solve military problems, and might not be sustained even by the new pro-defense Congress for more than two years. We discussed the possibilities of setting forth a new strategic approach and a technological end-run on the Soviets to meet President Reagan's commitment to a ''margin of safety.'' We decided to go to work in earnest to formulate such an approach.

The fundamental strategy change required was the replacement of the Mutual Assured Destruction (MAD) doctrine which had shaped—rather, *warped*—our strategic force posture and had undergirded the U.S. approach to arms control. The MAD doctrine postulates that strategic *defensive* systems are destabilizing and provocative, a theory that has led to a Free World seriously vulnerable to nuclear attack and blackmail.

Although military spokesmen had from time to time denied that MAD was U.S. policy, the political machinery adhered to it in essence. Certainly SALT negotiations had been conducted as if MAD were official U.S. strategy. SALT I attempted to disallow strategic defense by negating any significant antiballistic missile efforts. On the other hand, SALT I accommodated a massive offensive nuclear buildup then under way in the USSR and permitted a proliferation of city-busting MIRVs on our side.

U.S. negotiators later accepted a SALT II Treaty which was badly flawed in detail and which in general merely increased overall the limits on offensive nuclear power to accommodate quite obvious Soviet programs. Before the Senate rebelled and permitted SALT II to die, the standard rebuttal of Carter Administration spokesmen to critics of the Treaty was the MAD-based insistence that we would still be able to kill millions of Soviet civilians in a retaliatory strike with weapons we already have and could add more if we wished to do so under SALT II.

A search for technology which would provide the basis for an end-run on the Soviets led inexorably to space. The U.S. advantage in space is demonstrated in its most dramatic form by the Space Shuttle. More fundamentally, the ability of the United States to miniaturize components gives us great advantages in space where transport costs-per-pound are critical. Today, a pound of U.S. space machinery can do much more than a pound of Soviet space machinery.

It also happens that the technologies immediately available for military systems in space—beyond intelligence, communication, and navigation-aid satellites—are primarily applicable to ballistic missile defense systems. This fact raised a strong expectation that space held the key to a technological end-run which would offset current Soviet strategic nuclear advantages and at the same time provide an escape from the balance of terror doctrine of MAD.

Early in 1981, I wrote an article titled "Toward a New U.S. Strategy: Bold Strokes Rather than Increments," which was published in the Spring issue of *Strategic Review*. This article laid out the basic concept of a spaceborne defense which would nullify the MAD doctrine.

Although I was convinced that spaceborne defenses, perhaps using beam weapon technology (lasers, etc.), are feasible, I was unable to conceptualize a system which could stand up to doubters. However, in consultation with conceptual and technical experts working on other military space applications, we came up with a concept for a spaceborne ballistic missile defense system.

In order to avoid long lead times and interminable arguments among scientists, we sought to use already developed technology as much as possible. And in order to avoid a fruitless search for perfection, it was postulated that a system which could put at risk as little as 20 percent of an all-out missile attack on the United States would suffice, since even that modest level of attrition of a Soviet missile attack *in the early stages of trajectory* would be sufficient to destroy any confidence Moscow might have in a disarming first strike.

The solution we found was a spaceborne missile defense concept which can put at risk a much higher percentage of a Soviet strategic missile salvo fired not only at the U.S. but at our allies. The system concept, using off-the-shelf components, appears remarkably inexpensive and can probably be deployed in a relatively short time.

Thus far the global ballistic missile defense system concept postulated has held up well under severe scrutiny for feasibility, costs, timing, and vulnerability. *It may or may not be the best technical option available to us.* At a minimum, it has demonstrated the basic feasibility of spaceborne defenses which can fundamentally change the nature of the strategic balance away from Mutual Assured Destruction toward Assured Survival.

In the search for military options in space we were fortunate to rely on Brigadier General Robert Richardson (USAF Retired); the Honorable John Morse, former Deputy Assistant Secretary of Defense; and Arnold Kramish, our scientific advisor. Further, we are indebted to a group of Boeing Company engineers who scrubbed our results and provided invaluable advice.

Our horizons were expanded by Dr. Peter Glaser of Arthur D. Little Company, who convinced us that space held the key not only to national security but also to economic growth and energy supply. As a result of his input to our efforts, the High Frontier concept was broadened to constitute a true national strategy rather than a purely *military* strategy. We came to realize that military and nonmilitary efforts to tap U.S. opportunities in space would best proceed together in harness. Indeed, the attempt to separate these elements in government, and in public groups supporting space efforts, was a prime reason for lack of a vigorous, purposeful U.S. space effort. The Reagan Administration has taken laudable steps toward correcting this conceptual flaw.

In the Fall of 1981, High Frontier became a project of The Heritage Foundation where it has profited from the strong support of Mr. Edwin Feulner, Jr., President.

A great boost to the High Frontier concept was provided by Mr. Frank Barnett, who provided the opportunity to present it to an audience of distinguished citizens assembled in Washington by his National Strategy Information Center. The Honorable Karl R. Bendetsen was in the audience and spoke to me about his enthusiasm for the concepts. He contributed substantially to the ongoing momentum of the project and to definition and consensus.

This report has been compiled by the persons listed below. Each contributed to the effort not only in his area of expertise, but also to the development of the basic High Frontier concept:

Dr. Jeffrey Barlow
John Bosma
Philip C. Clarke
John J. Coakley
Dr. Miles Costick
Prof. James Dougherty
Dr. Jacquelyn K. Davis
Col. Sam Dickens, USAF (Ret.)
William J. Gill
Dr. Peter Glaser
Lawrence Hafstad
Dr. Mose Harvey
Frank Hoeber
Orlando Johnson
Cresson Kearny
Brig. Gen. Albion Knight, USA (Ret.)
Arnold Kramish
Cleveland Lane
Sherri Mayerhofer
Vicki McCowan
Marianne Mele
Maj. Gen. Stewart Meyer, USA (Ret.)
Ed Milauckas
Hon. John Morse
Dr. Ralph Nansen
Dr. Robert Pfaltzgraff, Jr.
Dr. Jerry Pournelle
John Rather
Fred W. Redding, Jr.
Brig. Gen. Robert Richardson, III, USAF (Ret.)
Dr. Peter Vajk
James Wilson

It is most gratifying that these people served and still serve High Frontier for no compensation or less compensation than they deserve or could receive elsewhere. This expression of concern for U.S. national interests and of support for the basic concept is heartening.

We are further indebted to the generosity of individuals and institutions whose major contributions made the High Frontier Study possible.

This report is the result of High Frontier thus far. There is much more to do before the promise becomes reality.

Daniel O. Graham
Washington, D.C.
February 28, 1982

THE HIGH FRONTIER STUDY: A SUMMARY

The United States is faced with an historic, but fleeting, opportunity to take its destiny into its own hands. The ominous military and economic trends which today beset the peoples of the Free World can be reversed, and confidence in the future of free political and economic systems can be restored.

To accomplish this, we need only take maximum advantage of one priceless legacy handed down to us by those free institutions—superiority in space technology. We can escape the brooding menace of "balance of terror" doctrines by deploying defensive systems in space. We can confound the prophets of doom by opening the vast and rich High Frontier of space for industrialization.

If we are to seize this historic opportunity, we must first muster the political will to discard without qualm the failed doctrines of the past, to attack without quarter the bureaucratic impediments to action, and to meet without flinching the wave of indignation from outraged ideologues at home and abroad. The technology is available, the costs are reasonable, and the alternatives are not promising solutions to our security problems.

THE OBJECTIVE

The objective of the High Frontier Study is to formulate a national strategy option which would make maximum use of U.S. space technology to accomplish the following goals:

- Nullify the present and growing threat to the U.S. and its allies which is posed by Soviet military power.
- Replace the dangerous doctrine of Mutual Assured Destruction (MAD) with a strategy of Assured Survival.
- Provide both security and incentive for realizing the enormous industrial and commercial potential of space.

GUIDELINES

This objective must be met with recommendations that are:

- Militarily sound,
- Technologically feasible,
- Fiscally responsible, and
- Politically practical.

THE THREAT IMPERATIVE

The High Frontier effort has focused primarily on countering the Soviet military threat which is ominous and growing. This threat is the result of determined efforts by the Soviet Union to establish global military dominance—efforts that have been abetted by poorly conceived U.S. security policies such as MAD. The Soviet military buildup coupled with U.S. military neglect has created these alarming conditions:

- There is a serious and growing Soviet advantage in strategic nuclear power which cannot be countered by the undefended United States except by a threat of retaliation that involves national suicide.
- The preponderance of Soviet conventional power vis-a-vis the U.S. and its allies is also severe and growing. It can no longer be counterbalanced, as it has been in the past, by a credible threat to bring higher technology U.S. weaponry to bear.
- The Soviet Union is increasingly successful in the use of propaganda and the application of direct or indirect military power to disrupt our alliances and to force the conversion of underdeveloped nations to Marxism. This

Soviet success now threatens the continuing availability of raw materials which are critical to the industrialized West.

- The West is dangerously dependent on diminishing crude oil supplies located in areas threatened by Soviet military or manipulative political power.
- The U.S. alliance system is in serious disarray. It suffers a lost sense of purpose and a perception of a decline in U.S. power and leadership. The Soviet propaganda offensive against U.S. nuclear weapons designed to persuade Europeans to become neutral is increasingly effective.

The Soviets are engaged in a costly and all too successful effort to cap their current strategic advantages—in their terms "a favorable correlation of forces"—with Soviet domination of near Earth space. The Soviets have the only tested space weapon on either side, an antisatellite system. They have orbited nuclear reactors. They have a manned space station in orbit and are expanding it. Almost all Soviet space activity has a distinct military flavor. The essence of the Soviet military space threat was included in the 1981 Department of Defense publication *Soviet Military Power* (pages 79-80):

> The Soviets have a vigorous and constantly expanding military space program. In the past ten years they have been launching spacecraft at over 75 per year, at the rate of four-to-five times that of the United States. The annual payload weight placed into orbit by the Soviets is even more impressive—660,000 pounds—ten times that of the United States. Some, but by no means all, of this differential can be accounted for by long-life U.S. satellites using miniaturized high technology components. Such an activity rate is expensive to underwrite, yet the Soviets are willing to expend resources on space hardware at an approximate eight percent per year growth rate in constant dollars.
>
> We estimate that 70 percent of Soviet space systems serve a purely military role, another 15 percent serve dual military/civil roles, and the remaining 15 percent are purely civil. The Soviet military satellites perform a wide variety of reconnaissance and collection missions. Military R&D experiments are performed onboard Soviet manned space stations, and the Soviets continue to develop and test an ASAT antisatellite co-orbital interceptor.
>
> The Soviets appear to be interested in and possibly developing an improved ASAT. A very large space booster similar in performance to the Apollo program's Saturn V is under development and will have the capability to launch very heavy payloads into orbit, including even larger and more capable laser weapons. This booster is estimated to have six-to-seven times the launch weight capability of the Space Shuttle.
>
> Soviet space research and development, test, production, and launch facilities are undergoing a continuing buildup. The new booster will be capable of putting very large permanently manned space stations into orbit. The Soviet goal of having continously manned space stations may support both defensive and offensive weapons in space with man in the space station for target selection, repairs and adjustments and positive command and control. The Soviet's predominantly military space program is expected to continue to produce steady gains in reliability, sophistication and operational capability.

The Soviets consider space a perfect environment in which to exercise their long standing doctrinal and operational preferences in warfighting—unconventional "first moves," preemptive attacks or "decapitation attacks" against vital targets such as strategic communications, "combined-arms" moves (as are possible with shiptracking satellites), and other elements of their well stocked repertoire. The Soviets integrate military space operations into their strategic thinking. They see space in straightforward terms, as an operational or combatant theater, whereas we see it—given our own strategic culture—as a "sanctuary" where "support forces" for terrestrial military forces can operate permissively.

If Moscow achieves its aims, we will be faced with a new era of *Pax Sovietica* in which Soviet space power dictates Free World behavior. We believe that the High Frontier of space provides us with the opportunity, perhaps our only opportunity, to frustrate Soviet power ambitions and at the same time open up a new era of hope and prosperity for the U.S. and the Free World.

THE HISTORICAL IMPERATIVE

The immediate threat impels us to exploit our space technology, but there is also an unavoidable historical imperative to move vigorously into that arena. Throughout man's history, those nations which moved most effectively from one arena of human activity to the next have reaped enormous strategic advantages. For instance, when man's activities moved from the land to the coastal seas, the Vikings established an extraordinary dominance by excelling at sailing those seas.

After the epic voyages of Columbus and Magellan, Spain and Portugal dominated the world through military and commercial control of the new arena of human activity—the high seas. Later England with her powerful fleet of merchantmen and men-of-war established a century of Pax Britannica. When the coastal seas of space—the air—became a new sphere of human activity, the United States gained great strategic advantages by acquiring the most effective military and civilian capability in aviation. Today, after epic manned and unmanned exploration of space, we shall see which nation puts the equivalent of the British merchantmen and men-of-war into space. We dare not let it be our adversary.

THE MILITARY DIMENSION

We cannot reverse the ominous trends in the military balance if we adhere to current strategy and try to compete with the Soviets in piling up weapons of current technology. Even if Congress were willing to appropriate unlimited funds for procurement of these weapons (and it is not), our defense production base is in such a sorry state that it could not compete with the Soviet arms production base which is today operating at very high levels. Our best hope is to change our strategy and to move the key competition into a technological arena where we have the advantage.

A bold and rapid entry into space, if announced and initiated now, would end-run the Soviets in the eyes of the world and move the contest into a new arena where we could exploit the technological advantages we hold. This is far preferable to pursuing a numbers contest here on Earth, which will be difficult if not impossible for us to win.

THE STRATEGIC DEFENSE OPTION

When we look to space for the technological end-run on the Soviets, we find all factors call for an emphasis on strategic defense. *First,* defensive systems hold the only promise to break out of the Mutual Assured Destruction doctrine. *Second,* defense is the only sound alternative to costly "racetrack"-type options to protect our deterrent systems. *Third,* our current and crucial heavy military investment in space is also vulnerable to attack. *Fourth,* available technology favors defensive space systems. *Last,* there are severe political constraints and some technical-military reasons inhibiting the deployment of offensive weapons in space.

For these reasons the military side of High Frontier emphasizes the resurrection of a long neglected aspect of our security—protective strategic defense. We visualize a layered strategic defense. The first layer would be a spaceborne defense which would effectively filter a Soviet missile attack in the early stages of flight. The second layer would be a broader space protection system, perhaps using advanced beam weaponry to further reduce the effectiveness of a missile attack and to defend other space assets from a variety of attacks. The third layer would be a ground based point defense system capable of removing any Soviet assurance of success of a first strike against our missile silos—even *before* a space system is deployed—and of intercepting Soviet missiles which later might leak through the space defenses. A passive fourth layer would be civil defense, which becomes a valuable aspect of strategy in conjunction with these active defense layers.

We can get a point defense within two or three years which would be adequate to protect our ICBMs in silos and avoid the high cost deployment modes for MX. An initial spaceborne global ballistic missile defense (GBMD) can be acquired in five or six years given adequate priority. A second generation general space defense using more advanced technology can probably be achieved in the early 1990s.

In proposing such strategic defenses, one invariably encounters the shibboleths that have plagued consideration of strategic defensive options in the past. It has been an article of faith in the offense-only, Assured Destruction school of thought that strategic defenses in the nuclear era are useless unless they are impermeable or not subject to attack and/or that they are impossibly expensive. These are false premises.

With regard to impermeable or invulnerable defenses, there never has been nor ever will be a defensive system which could meet such criteria. Such perfectionist demands ignore the purposes of defenses and the effects of strategic defense on deterrence. Defenses throughout military history have been designed to make attack more difficult and more costly—not *impossible.* Defenses have often prevented attack by making its outcome uncertain. General Grant put a cavalry screen in front of his forces not because the cavalry was invulnerable to Confederate bullets or because he thought it could defeat General Lee, but because he did not want the battle to commence with an assault on his main forces or his headquarters.

It is this same military common sense that must prevail in our approach to strategic defenses today. Given the drastic consequences of a failed nuclear attack on an opponent, the critical military task is to keep a potential aggressor *uncertain of success,* if not certain of failure. In the absence of defenses, the Soviet military planner has a rather straightforward arithmetic problem to solve to be quite sure of the results of a disarming strike against all locatable U.S. strategic weaponry—ICBM sites, airfields, and submarine bases. His problem is simply to ensure that he can deliver two warheads of current size and accuracy against each such target. If, on the other hand, the Soviet planner must consider the effects of a strategic defense, especially a spaceborne defense which destroys a portion of the attacking missiles in the early stages of their trajectories, he is faced with a problem full of uncertainties. He does not know how many warheads will arrive in the target area and—even more crucial—which *ones* will arrive over which targets. This changes the simple arithmetic problem into a complex calculus full of uncertainties. *Such uncertainties are the essence of deterrence.*

Strategic defenses are eminently practicable and by no means impossibly expensive if the programs involved are not required to meet unrealistic standards of perfection or incredible postulated threats. A cursory review of combinations of spaceborne defenses, land based ABMs, and civil defense—while by no means *definitive* as to costs—indicates that a defense system of decisive strategic importance can be devised which is relatively inexpensive when compared with some previously proposed offensive systems.

SURVIVABILITY

One issue which must be carefully addressed is that of space system survivability. While space systems are nearly invulnerable to a large array of threats with which terrestrial systems must cope (e.g., bombs and bullets), they have some unique vulnerabilities to threats which can be posed by a technologically advanced adversary. An examination of this problem leads to several conclusions:

- As with all other systems, no space based system can be envisaged which is invulnerable to *all* postulated threats.
- Vulnerability of current U.S. space assets (intelligence and communications satellites and the Shuttle) sharply increases the imperative for an effective spaceborne defensive system which can defend itself and reduce the threat to other space systems, as well as defend ground targets against hostile objects transiting space, e.g. ICBMs.
- Defensive systems employing large numbers of less sophisticated satellites are far less vulnerable than those employing small numbers of more sophisticated satellites.
- An ability to provide mutual warning and protection among satellites in a ballistic missile defense is very important to survivability.
- The sooner a spaceborne ballistic missile defense system can be deployed, the better its survivability (long lead time systems are susceptible to long lead time Soviet countermeasures—real or postulated).
- Future U.S. deployment of more sophisticated beam weapon military satellites may be dependent for survivability on protection provided by a lower technology defensive system already deployed.

Given the characteristics of currently operating U.S. space systems, one can readily postulate ways for the Soviets to attack them, ranging all the way from throwing sand in their paths to burning them out of space with futuristic beam weapons. Such attack modes fall into two basic categories, peacetime attack and wartime attack.

Most *current* Soviet capabilities to attack U.S. space systems are applicable in the peacetime attack category. These include attack with non-nuclear direct ascent missiles, the current Soviet antisatellite system, and current power level Soviet lasers. However, these attack modes presuppose Soviet willingness to risk the grave consequences (including war) of attacking our space systems in time of peace or crisis. While such Soviet action cannot be totally ignored, most experts on Soviet behavior find this possibility extremely remote.

The second class of threat—wartime—is more serious. In this situation nuclear weapons could be used to destroy or disable our space systems using radiation effects. (Blast effects are of little effect outside the atmosphere.) There are technical means of reducing the vulnerability of space systems to these effects, but a capability of a defensive system to intercept hostile objects directed at it is the best counter to such threats.

The Soviets may develop laser beam weaponry of such power that satellites passing over them could be destroyed with a single burst of energy. It is doubtful, however, that such systems could, in the foreseeable future, successfully attack satellites coming over the horizon toward the Soviet Union where they would be shielded by much more of the Earth's atmosphere.

Probably the most important factors in the survivability problem are military rather than technical. Survivability is sharply increased by the ability of space vehicles to destroy threatening objects launched at them, or at other U.S. space vehicles. Even if the Soviets eventually create the means to attack a spaceborne defense system successfully in order to launch a strategic missile attack subsequently, all chances of destroying the U.S. deterrent on the ground would be lost. In these circumstances, launch on warning or launch under attach become both credible and feasible options for the United States. The Soviets could not expect, after the attack in space, that the U.S. President would hesitate to respond to sensor warnings that a missile attack had been launched from the USSR. This fact alone would make a spaceborne defense of great strategic value.

NONMILITARY DIMENSION

Space holds out the promise of a new era of economic expansion. The unique environment of space—zero gravity, near perfect vacuum, unlimited heat absorption, and sterile conditions—opens up a broad range of industrial/commercial possibilities. Space also contains inexhaustible supplies of minerals and solar energy. The economic potential of space is already being tapped in the communications industry. As the cost of space transportation is lowered, the industrialization of space will burgeon. However, the capital investment in space industries will be quite large and unlikely to be undertaken if space installations are indefensible against hostile attack. For this reason, military capabilities in space are critical to space based economic growth.

We should harbor no illusions that space can be limited to "peaceful uses" any more than could previous arenas on land, sea, or in the air. Indeed, most current space assets, U.S. and Soviet, are partially or entirely military—and the most destructive of all weapons, strategic ballistic missiles, must transit space en route to their targets.

The government's role in opening up the High Frontier of space for economic exploitation is basically the same as it has been with the opening of frontiers of the past—exploration, transportation systems, and security. These functions translate to these specifics: scientific research, improving the Space Shuttle, and providing spaceborne defenses.

Both the military and nonmilitary uses of space depend on the continued efforts in certain core technologies: improvements in space transportation to reduce the cost-per-pound of materials in orbit, and the creation of permanent, manned space stations at the "terminals" of the space transport system.

While these efforts are primarily the responsibility of government, they should be undertaken in cooperation with private industry and with support from other nations which would benefit.

With a proper combination of space technologies, we can sharply improve the security of the U.S. and its Free World allies and, at the same time, restore confidence in the ability of Free World economies to meet the challenges of the future.

The urgency here is far greater than many people in this country appear to recognize. Following the successful U.S. Moon landing, the Soviets made it clear that, while intending first and foremost to develop maximum possible military capabilities in space, they expect also to achieve dominance with respect to the economic exploitation of space opportunities. In 1964 Brezhnev spoke of these plans, and Soviet specialized literature has gone into great detail concerning concrete possibilities. Further, all phases of ongoing Soviet space activities that aim at strategic objectives also serve as stepping-stones to the USSR's preeminence in the space environment for military as well as nonmilitary purposes.

THE URGENT REQUIREMENTS

In order to fulfill the objectives of the High Frontier concept, including the rapid closing of the "window of vulnerability," creating the concrete basis for a new strategy of Assured Survival, and opening space for economic growth, the following list of urgent requirements is presented. It should be noted that these requirements, when met, will not solve *all* urgent military problems facing the United States, let alone all economic problems.

The urgent requirements for military systems to implement the High Frontier concept are these:

1. A point defense for U.S. ICBM silos which, within two or three years, at a cost less than that of superhardening, can destroy any confidence the Soviets might have in a first strike against our deterrent.
2. A first generation spaceborne ballistic missile defense, deployable in five or six years at a cost not exceeding that of the original MX-MPS system, and capable of significant attrition of a Soviet strategic missile attack in the early part of trajectory.
3. A second generation space defense system, deployable within 10 or 12 years and capable of attacking hostile objects anywhere in near Earth space with advanced technology weaponry.
4. A utilitarian manned military space control vehicle, deployable within the next six to eight years, and capable of inspection, on-orbit maintenance and space tug missions wherever satellites can go.
5. A civil defense program of sufficient scope and funding to take advantage of the proposed active strategic defenses and thus add to U.S. deterrent strength.

The primary urgent requirements in core space technology and nonmilitary applications are:

1. Improved space transportation, designed to lower the cost-per-pound in orbit to under $100.
2. A manned space station in low Earth orbit as soon as practicable. It would allow low cost, efficient development and testing of both civilian and military system elements, and constitute a first step toward a similar manned station at geosynchronous orbit.
3. Development work on reliable, high capacity energy systems in space, initially to power other space activities, and eventually to provide electrical power to any spot on Earth.
4. Preparatory development of a selected number of promising commercial business opportunities. Government efforts should focus on encouraging the transformation of these "seed" efforts into independently viable commercial operations as soon as possible.

CAN WE DO IT?

All these requirements can be met, some of them with technology already in hand, with components already tested. None of these requirements demand technological "breakthroughs" or a commitment to mere scientific theories. There are in fact a variety of viable options available to meet each of the requirements of High Frontier. The following is a description of one set of programs which could do so. Each is described in some detail in the main body of this study. The costs estimated for these programs are in constant dollars. The costs and times indicated are based on a management system which minimizes bureaucratic delays.

QUICKLY DEPLOYABLE POINT DEFENSE

A partially tested system exists that could meet the requirement to destroy Soviet confidence in a first strike against our silos. It is a very simple system which fires a large number of small conventional projectiles which form a barrier against a warhead approaching a U.S. missile silo at about one mile from the target. It could be described as "dynamic hardening" instead of as an antimissile system. If deployed to intercept only the first Soviet warhead approaching a silo, it would cost $2-3 million per defended silo. If it is to intercept a second warhead, the costs increase to about $5 million per silo.

FIRST GENERATION SPACEBORNE DEFENSE

The requirement for an initial spaceborne ballistic missile defense system can be met by using off-the-shelf hardware to create a multiple vehicle, orbiting system. This system would deploy non-nuclear kill vehicles to destroy Soviet missiles in the early phase of trajectory. Enough weapons carrying satellites would be orbited to ensure continuous coverage of Soviet ballistic missile trajectories, including those of SS-20 Eurostrategic missiles and submarine launched missiles. This system could provide protection to the allies as well as to the United States.

The multiple satellite deployment permits one satellite to defend itself and several others from hostile attack. It also has the potential for forming the basis of a highly effective and secure command, control, and communications (C^3) system. Since the system makes maximum use of off-the-shelf space hardware components, it may be the cheapest and quickest available option. This system could start deployment in perhaps as little as three years and be fully deployed in five or six years at a minimum cost of some $10-15 billion.

SECOND GENERATION SPACEBORNE DEFENSE

The most promising possibility for a second generation spaceborne defense is product improvement of GBMD I. With the addition of advanced infrared sensing devices the first generation can be made capable of attacking individual warheads throughout their trajectory up to reentry into the atmosphere. This system could be ready for deployment in 1990 at a cost of about a $5 billion add-on to GBMD I costs.

The requirement for higher technology space defense systems might also be met by a high powered laser system on the ground with redirecting mirrors on satellites or by beam weapon systems deployed in space or in pop-up installations on the ground. These systems are currently being researched. Costs to continue research should probably be increased by about $100 million per year.

HIGH PERFORMANCE SPACEPLANE

There is an urgent need to develop a multipurpose, military, manned space control vehicle to perform a wide variety of space missions such as inspection of friendly or suspect space objects, satellite and space station protection, and adjustment or retrieval of satellites. One such vehicle is the high performance spaceplane, or one man "space cruiser," which utilizes available space hardware components and technology and which could be operating in several years for less than $500 million in cost. It is now under active consideration in the Department of Defense.

CIVIL DEFENSE

Civil defense is a multifaceted endeavor, the utility and cost effectiveness of which sharply increase when considered in conjunction with active defenses. This study concludes that increased funding for civil defense is required for the near term but that over the longer term the active defenses of High Frontier would reduce the requirement for resource expenditures on civil defense. The impact of these conclusions on priorities and costs of current civil defense programs has not been analyzed in this study.

IMPROVED SPACE TRANSPORTATION

The immediate answer to improved space transportation is an upgrade of the current Shuttle program to improve turnaround time and to create an unmanned cargo-only version. At the same time, development work should begin on a much higher load capacity vehicle. These programs would cost an estimated $6 billion over a 10-year period.

A MANNED LOW EARTH ORBIT SPACE STATION

The currently proposed military Space Operation Center should be given high priority and expanded in concept to include provision for "fly-along" industrial/commercial space installations. The space station should be equipped to receive power for operations from a prototype solar power satellite. A 10-year program to deploy this space station should cost about $12 billion.

A SPACE POWER SYSTEM

This requirement can be met by a proposal using known technology which would place in geosynchronous orbit a solar power satellite and place on Earth a microwave receiving antenna and conversion system providing 500 megawatts of continuous electrical power. This pilot system, modified to include a capability to provide power to a space station with laser transmission, would cost about $13 billion.

SPACE INDUSTRIAL SYSTEMS RESEARCH AND DEVELOPMENT

The costs of R&D for industrial space applications would probably be borne almost entirely by interested private enterprise, with no more than $50 million per year in government support.

COSTS

The total costs of the High Frontier concept over the next five or six years in outlays of constant dollars might be on the order of $24 billion. Through 1990 the total costs in constant dollars would probably be about $40 billion—a figure that compares favorably with what would have been the total cost of MX-MPS in its original configuration. It also compares favorably with the Apollo Moon-landing program, and strikingly so if the inflation rate of the past 12 years is considered.

If one considers possible tradeoffs in programs no longer needed or lowered in priority by the existence of an effective strategic defense, the real costs of the High Frontier programs are even lower. For instance, the billions now earmarked for superhardening of existing missile silos and for deploying more complex point defenses need not be expended. There are other possible tradeoffs such as repositioning of SAC airfields, reducing the urgency of theater nuclear force upgrade in Europe, C^3 improvements, and so forth.

Finally, there is a reasonable chance for sizeable cost offsets from industry and allied participation in the most expensive aspects of the High Frontier effort—nonmilitary applications. This is especially true if a vigorous effort to tap solar energy is emphasized. Several nations have already stated their willingness to assist in such an effort. Such nongovernment support would further reduce the real costs of the concept.

In any case, costs to the U.S. taxpayer of implementing High Frontier will certainly be lower than those involved in other approaches to solving urgent security issues, e.g., MX-MPS. The High Frontier approach, therefore, cannot be characterized as unrealistically expensive.

IMPACTS

The mere announcement of a bold, new U.S. initiative along the lines of the High Frontier concept would have beneficial impacts at home and abroad. The *fulfillment* of the urgent requirements noted above would have even more far reaching impacts.

MILITARY IMPACTS

On the purely military-strategic side, we would be moving away from the unstable world of terror balance to one of Assured Survival—a much more stable condition. We would provide answers to U.S. and allied security problems not involving the amassing of ever larger stockpiles and ever more expensive deployments of nuclear weapons.

By creating a proper balance between strategic offense and strategic defense we broaden the options for strategic retaliatory systems. A great deal of the counterforce, damage-limiting function of our strategic forces can be shouldered by the defensive systems. Cruise missiles become a more attractive option in a new strategic setting that includes defenses against ballistic missile attack.

Perhaps most important to our military efforts as a whole, the High Frontier concept would restore the traditional U.S. military ethic. The military man's role as defender of the country has always been the tie that has bound him to the supporting citizenry. Strategies of the recent past, such as MAD, which deny that role have seriously weakened that bond. A commitment to a new strategy which is consistent with the military rationale of the average U.S. citizen could greatly ease problems in all facets of U.S. security efforts.

POLITICAL IMPACTS

The potential for public support of this concept is enormous. If the military and nonmilitary aspects of High Frontier are effectively harnessed together, broad segments of the U.S. body politic are likely to rally in support. Recent elections have demonstrated the widespread desire for improved defenses. There is a remarkably large support base, primarily among younger people, in the form of space enthusiasts. And there is general public disillusionment with the doctrines and strategies of the past.

The High Frontier concept would even convert or confuse some of the conventional opponents of defense efforts and technological innovations. It is harder to oppose nonnuclear defensive systems than nuclear offensive systems. It is impossible to argue effectively for a perpetual balance of terror if it can be negated by new policies. It is hard to make environmentalist cases against space systems.

Even those naysayers whose basic concern is disarmament will be hard pressed to make a case against High Frontier, the ABM Treaty notwithstanding. It is not necessary to abrogate the ABM Treaty to commit to High Frontier programs.

The High Frontier spaceborne defensive systems fall into the category described in the treaty as "systems based on other principles" which are "subject to discussion" with the Soviets. Point defense systems can be selected which are so different from ABM systems as defined in the treaty, that they too could be considered as outside the treaty. Indeed, some silo defense systems can be considered "dynamic hardening"—a substitute for reinforced concrete—rather than an ABM. Further, the current ABM Treaty is scheduled for review in 1982, and the United States could propose any amendments deemed necessary to accommodate strategic defensive decisions.

A U.S. commitment to the High Frontier concept does not necessitate rejection of arms negotiations with the Soviets. It does, however, mean that future negotiations would proceed on a different philosophical basis. Rather than continue to pursue agreements which attempt to perpetuate a balance of terror and MAD, our negotiating efforts would be dedicated to achieving a stable world of Mutual Assured Survival.

ECONOMIC IMPACTS

There can be little doubt that a strong commitment by the United States would have highly beneficial economic impacts. Some of these impacts will affect the U.S. economy in the near term, primarily through the stimulus to investment in high technology sectors of industry and a probable upswing in confidence generally. An increase of 200,000 jobs in the near term as a result of a strong commitment to space has been estimated. Longer term impacts will depend on the rate at which industrial applications are realized and on unpredictable technological spin-offs from the space effort.

One area of commercial space application is already paying its way very well. Space communications is a $500 million-per-year enterprise and is growing rapidly. By 1990 it should become a multibillion dollar-per-year industry.

As other industrial applications in space are realized, the total revenues from space industries might reach levels of several tens of billion dollars-per-year by the year 2000.

Some of the most beneficial economic impacts of a strong High Frontier effort are indirect and unquantifiable. The demand for highly skilled workers is certain to have an impact on the education system and on the labor market. New products, tools, and services will be required by an expanding space effort. Research efforts will intensify.

Overall, the economic benefits of a strong U.S. commitment to the exploitation of space for both security and industry are potentially very great, but they are no more predictable today than were the future economic benefits of aviation in the 1920s.

FOREIGN IMPACTS

The positive political effects in the U.S. will probably be reflected overseas among our allies. The announcement of a commitment to the High Frontier concepts could have a strong countereffect on the current highly disruptive, "antinuclear," or "peace" movements in Europe. A bold U.S. strategic initiative would certainly bolster the morale of pro-U.S. elements. The High Frontier concept can become a new cement for Free World alliances, making them global rather than regional.

A shared U.S.-Allied commitment to the harnessing of solar power from space could have highly beneficial impacts on foreign relations. If the prospects were good for future supplies of energy independent of the geographical location of fossil fuels, the overdependence of the industrialized West on oil and gas producing countries could be rectified. Further, the prospects for overcoming the intractable problems of the underdeveloped nations could have a beneficial impact on the attitudes of the Third World.

As for the Soviets, their reaction is easily predictable as hostile. They have already moved to counter the U.S. potential to adopt available military space options. They have introduced in the United Nations (and garnered some support for it among our allies) a new treaty which would ban *all* (not just nuclear) weapons in space. Meanwhile, evidence mounts that they are already in violation of their own cynical proposition. We can expect an extraordinarily strong Soviet propaganda effort against a U.S. commitment to the High Frontier concepts, including threats of counteraction. However, in both particulars Moscow will find, for substantive reasons, an attack on the High Frontier concepts much more difficult to conduct than past anti-U.S. campaigns.

MANAGEMENT

Time is critical in any commitment to the High Frontier, especially with regard to the military systems. If we cannot change the adverse trends in the military balance quickly, we may not be able to change them at all. If we do not move quickly to secure space for promising industrial development, we may later be denied the opportunity.

There are no technical obstacles to meeting the military and nonmilitary objectives of High Frontier. We can close the window of vulnerability in two or three years and negate the brooding menace of Mutual Assured Destruction in five or six years. We can lower the costs of men and materials in space, establish a permanent manned presence in space, and open the door to enormous economic advantages in 10 years. However, this can be done only by initially selecting systems using off-the-shelf technology to the maximum and by instituting special management and procedural arrangements for their rapid acquisition and deployment. By using known and tested technology we can avoid the long delays imposed by research and development. By special management arrangements we can avoid the bureaucratic hurdles which have been inserted into our weapons acquisition processes over the past 15 years. Time is money, and literally billions can be saved by cutting acquisition times.

In 1956, President Eisenhower gave the go-ahead on a concept for a ballistic missile firing submarine. That concept involved far more technological unknowns than do the High Frontier options. In 1960, 47 months later, the first Polaris put to sea. In 1962, President Kennedy announced the objective of landing a man on the Moon. Seven years later this astonishing feat was accomplished.

Today, even a new fighter aircraft takes 13 years or more from concept to acquisition, and decades of delay are predicted for space developments. Such protracted processes cause costs to soar astronomically. This sad state of affairs exists not because Americans have become technologically inept but because we have, over the years, constructed a complex and multilayered bureaucratic system in the Executive Branch and in the Congress which simply cannot produce quick results. In order to take advantage of the opportunities available to us on the High Frontier, we must—at least for a few years—find a way to short circuit the bureaucratic institutions and procedures.

The first step is to select—and select quickly—those systems which will meet the urgent requirements of the High Frontier concept. This should be done by a Presidential Systems Selection Task Force composed of prominent and properly qualified individuals.

To provide overall guidance to the High Frontier effort, a National Space Council should be appointed with representation from the involved departments and agencies of the Executive Branch, the Congress, and industry. Its function would be to insure full cooperation and fast action by all branches of government and of private industry involved in the effort. Its chairman should be the Vice President.

The actual coordinating and expediting of the programs selected to meet the High Frontier requirements should be the responsibility of a chief operating officer heading up a Consolidated Program Office. This officer should be assisted by special project officers within the departments and agencies charged with acquiring the first generation of High Frontier systems. The management system should insure individual rather than committee responsibility for decisions, a minimum of Executive and Congressional staff review, and specified or ''fenced'' funding for High Frontier programs.

This management system should be unequivocally temporary. It should go out of existence upon achievement of its objectives of first generation system acquisition. As results are obtained, all responsibility for the operations, maintenance, and further growth of space systems should return to the cognizance of the appropriate agencies—Defense and NASA. There is no need to create a new permanent layer of bureaucracy.

These are the essentials of the High Frontier concept. They are discussed in much greater detail in the main body of the study. We believe that the change of strategy recommended in this study supports a U.S. policy statement as follows:

PROPOSED STATEMENT OF U.S. POLICY

The United States and its allies now have the combined technological, economic, and moral means to overcome many of the ills that beset our civilization. We need not pass on to our children the horrendous legacy of "Mutual Assured Destruction," a perpetual balance of terror that can but favor those most inclined to use terror to bring down our free societies. We need not succumb to ever gloomier predictions of diminishing energy, raw materials, and food supplies. We need not resign ourselves to a constant retreat of free economic and political systems in the face of totalitarian aggressions. The peoples of the Free World can once again take charge of their destinies, if they but muster the will to do so.

In April of 1981, the Space Shuttle Columbia made its dramatic maiden voyage into space and back safely to Earth. This event was not merely another admirable feat of American space technology. It marked the advent of a new era of human activity on the High Frontier of space. The Space Shuttle is a development even more momentous for the future of mankind than was the completion of the transcontinental railway, the Suez and Panama Canals, or the first flight of the Wright brothers. It can be viewed as a "railroad into space" over which will move the men and materials necessary to open broad new fields of human endeavor in space and to free us from the brooding menace of nuclear attack.

This is an historic opportunity—history is driving us to seize it.

A few thousand years ago, man's activities—his work, his commerce, his communications, all of his activities, including armed conflict—were confined to the land.

Eventually man's technology and daring thrust his activities off the land areas of the continents and into the coastal seas. His work, commerce, communications, and military capabilities moved strongly into this new arena of human activity. Those nations that had either the wit or the luck to establish the strongest military and commercial capabilities in the new arena reaped enormous strategic advantages. For example, the Vikings, although never a very numerous people, became such masters of the coastal seas that their power spread from their homes in Scandinavia over all the coasts of Europe and into the Mediterranean Sea, up to the very gates of Byzantium.

At the beginning of the 16th century, after the epic voyages of men like Magellan and Columbus, human activity surged onto the high seas. Once again, the nations that mastered this new arena of human activity reaped enormous strategic rewards. First Spain and Portugal utilized their sea power to found colonies and to solidify their strength in Europe. Later, Great Britain, with an unsurpassed fleet of merchantmen and fighting ships, established a century of relative peace which we remember as Pax Britannica.

In the lifetime of many of us, man's activity moved strongly into yet another arena, the coastal seas of space—the air. And once again the nations which quickly and effectively made use of this new arena for commerce and defense gained great advantages. As Americans we can take pride that the greatest commercial and military successes in aviation have been achieved by our nation.

But today, following the epic voyages of our astronauts to the Moon and our unmanned explorer satellites to the rings of Saturn and beyond, we find man's activities moving strongly into yet another new arena—the high seas of space. Already the United States and other major nations, including the Soviet Union, are making huge investments in space. Much of our communications, intelligence, weather forecasting, and navigation capabilities are now heavily dependent on space satellites. And, as history teaches us well, those nations or groups of nations that become preeminent in space will gain the decisive advantage of this strategic "high ground."

We must be determined that these advantages shall accrue to the peoples of the Free World; not to any totalitarian power. We can improve the Shuttle, our railway into space, placing space stations at its terminals and sharply reducing the cost-per-pound of material put into space. We can thus open the doors of opportunity to develop entire new space based industries, promising new products and new jobs for our people on Earth. We can eventually create the means to bring back to Earth the minerals and the inexhaustible solar energy available in space. By doing so, we can confound the gloomy predictions of diminishing energy and material resources available here on Earth. This will not only enhance the prosperity of the advanced, industrialized nations of our Free World, but will also provide the means to solve many of the hitherto intractable problems of the developing countries.

Further, we can place into space the means to defend these peaceful endeavors from interference or attack by any hostile power. We can deploy in space a purely defensive system of satellites using nonnuclear weapons which will deny any hostile power a rational option for attacking our current and future space vehicles or for delivering a militarily effective first strike, with its strategic ballistic missiles on our country or on the territory of our allies. Such a global ballistic missile defense system is well within our present technological capabilities and can be deployed in space in this decade, at less cost than other options that might be available to us to redress the strategic balance.

We need not abrogate current treaties to pursue these defensive options. A United Nations Treaty prohibits the emplacement of weapons of mass destruction in space, but does not prohibit defensive space weapons. The ABM Treaty requires discussion among Soviet and U.S. representatives of any decision to proceed with defensive systems "based on other principles" such as space systems. We should initiate such discussions and propose revisions, if necessary, in the ABM Treaty which is scheduled for review in 1982.

Essentially, this is a decision to provide an effective defense against nuclear attack for our country and our allies. It represents a long overdue concrete rejection by this country of the "Mutual Assured Destruction" theory which held that the only effective deterrent to nuclear war was a permanent threat by the United States and the Soviet Union to heap nuclear devastation on the cities and populations of each other. The inescapable corollary of this theory of MAD (perhaps the most apt acronym ever devised in Washington) was that civilian populations should not be defended, as they were to be considered hostages in this monstrous balance of terror doctrine. The MAD doctrine, which holds that attempting to defend ourselves would be "destabilizing" and "provocative," has resulted not only in the neglect of our active military and strategic defenses and our civil defense, it has also resulted in the near total dismantlement of such strategic defenses as we once had.

For years, many of our top military men have decried the devastating effect the MAD theory has had on the nation's security. In fact, our military leaders have, over the years, denied its validity and tried within the limits of their prerogatives to offset its ill effects. But those effects are readily evident. The only response permitted under MAD to increased nuclear threats to the United States or to its allies was to match these threats with increased nuclear threats against the Soviet Union. Further, a U.S. strategy which relied at its core on the capability to annihilate civilians and denied the soldier his traditional role of defending his fellow citizens has had a deleterious effect on the traditional American military ethic, and on the relationship between the soldier and the normally highly supportive public.

This legacy of MAD lies at the heart of many current problems of U.S. and allied security. We should abandon this immoral and militarily bankrupt theory of MAD and move from "Mutual Assured Destruction" to "Assured Survival." Should the Soviet Union wish to join in this endeavor—to make Assured Survival a mutual endeavor—we would, of course, not object. We have an abiding and vital interest in assuring the survival of our nation and our allies. We have no interest in the nuclear devastation of the Soviet Union.

If both East and West can free themselves from the threat of disarming nuclear first strikes, both sides will have little compulsion to amass ever larger arsenals of nuclear weapons. This would most certainly produce a more peaceful and stable world than the one we now inhabit. And it would allow us to avoid leaving to future generations the horrendous legacy of a perpetual balance of terror.

What we propose is not a panacea which solves all the problems of our national security. Spaceborne defense does not mean that our nuclear retaliatory capabilities can be abandoned or neglected. The United States would still maintain strategic offensive forces capable of retaliation in case of attack. The Soviets, while losing their advantage in first strike capabilities, would still be able to retaliate in case of attack. Nor does our approach to the strategic nuclear balance eliminate the need to build and maintain strong conventional capabilities.

We Americans have always been successful on the frontiers; we will be successful on the new High Frontier of space. We need only be as bold and resourceful as our forefathers.

CHAPTER I: STRATEGY

The participants in the High Frontier Study have intensively reviewed the spectrum of threats facing the United States and its allies, the global problems associated with those threats, and the options available to meet them. We urgently recommend the adoption of a new national strategy of Assured Survival to replace Mutual Assured Destruction. Assured Survival can be achieved by using U.S. technological advantages, especially in space, to provide our citizens with long neglected protection from nuclear attack and to secure space for our long term economic benefit.

This new strategy would:

- Provide for the defense of the United States and its allies against nuclear ballistic missile attack.
- Secure the availability of the vast resources of space to the United States and to the Free World by providing for its defense against hostile attempts to deny the use of this great medium for peaceful purposes.
- Capture the imagination and support of the broadest spectrum of peoples of the Free World and restore to them a sense of optimism and purpose through neutralizing the Soviet's strategic nuclear menace.
- Require the vigorous development of the economic opportunities available to us on the High Frontier of space—opportunities open to us because of hard won advantages in space technology—for the benefit and prosperity of the industrial nations of the Free World and to address the presently intractable problems of the lesser developed nations.

We believe that such a strategy is sound, technologically feasible, well within our capabilities, fiscally within our means, and likely to engender strong public support at home and abroad.

THE SITUATION IN BRIEF

More than a decade of adherence to unsound policies and doctrines based on illusory notions concerning the nature of threats to U.S. security and vital interests has led to a situation in which:

- There is a serious and growing Soviet advantage in strategic power which cannot be countered by the undefended United States except by a threat of retaliation that involves national suicide.
- The preponderance of Soviet conventional power vis-a-vis the U.S. and its allies is also severe and growing. It can no longer be counterbalanced, as it has been in the past, by a credible threat to bring higher technology U.S. weaponry to bear.
- The Soviet Union is increasingly successful in the use of propaganda and the application of direct or indirect military power to disrupt our alliances and to force the conversion of underdeveloped nations to Marxism. This Soviet success now threatens the continuing availability of raw materials which are critical to the industrialized West.
- The West is dangerously dependent on diminishing crude oil supplies located in areas threatened by Soviet military or manipulative political power.
- The U.S. alliance system is in serious disarray. It suffers a lost sense of purpose and a perception of a decline in U.S. power and leadership. The Soviet propaganda offensive against U.S. nuclear weapons designed to

persuade Europeans to become neutral is increasingly effective.

This litany of Western woes is not the whole picture, however. The USSR has problems of its own. It is suffering the strains of imperium in Poland and Afghanistan. Inside the USSR, the Kremlin is faced with a small but growing group of dissidents among its elites and the future prospects of Great Russians becoming an ethnic minority. Further, the Soviet Union's huge commitment of resources to its military machine over the past 15 years has impaired its already chronically deficient general economy.

This combination of threat and opportunity provides the United States with an historic but fleeting opportunity to change the world for the better.

THE OPTIONS

There are two basic options available to meet the military strategic challenge: an incremental approach, or a bold new initiative.

THE INCREMENTAL APPROACH

We could attempt to address the situation by merely modifying the basic strategies of the past and adding resources to the programs designed to support that strategy. This would entail incremental changes in our strategy and in our military programs along these lines:

- Mutual Assured Destruction would remain the unspoken (although frequently *denied*) cornerstone of our force structure, *but* we would modify our offensive forces to insure a somewhat higher level of destruction to the USSR in the event of attack, while continuing to eschew active strategic defense or effective civil defense.
- We would continue to rely on arms control treaties (past and future) as the answer to national security at reasonable cost, *but* we would get tougher with the Soviets at the bargaining table.
- We would remain content with the concepts of "parity," "essential equivalence," etc., *but* we would decry numerical imbalances and add billions of dollars to military programs presumed to close gaps between the U.S. and the Soviets in current technology weapons and forces.

THE BOLD APPROACH

The other basic option envisions a new national strategy, rejects the MAD doctrine outright, faces squarely the failure of arms control efforts to date, and end-runs current Soviet strategic advantages with superior technology. This bold approach entails:

- Replacing Mutual Assured Destruction with a strategy of Assured Survival, through an emphasis on strategic defense which maximizes the use of already known space technology and available point defense options. This would create, in the shortest possible time, an effective Free World defense against Soviet nuclear attack or blackmail.
- Adopting systems which would be readily adaptable over the years to foreseeable technological advances, such as laser and other beam weaponry.
- Attacking the broader spectrum of national problems by pursuing parallel nonmilitary programs to open and defend space for promising industrial and commercial development.
- Initiating near term nonspace programs which would be compatible with the new strategy and which would enhance the value of mid and longer term space programs.
- Holding total costs at or below current budget projections.
- Creating a management structure which would overcome the long lead times inherent in new systems development and acquisition procedures.

A COMPARISON OF THE OPTIONS

The incremental approach will almost certainly fail economically, strategically, and politically because:

- Attempts to close the arms gap with the Soviets by adding more hardware of current technology (missiles, aircraft, ships, tanks, etc.) play to their long suit. The Soviets are already producing these items at a very high rate, far surpassing current production rates of the U.S. and its allies. With a universal draft, they are also able to man that hardware effectively. In a contest for sheer military mass, the Soviets will probably be even further ahead of us in four years than they are today (see Figure 1).
- Our industrial mobilization base is grossly inadequate and in reality is incapable of closing the gap in hardware within an acceptable timeframe.
- Mere modifications to previous strategy (MAD, disarmament, detente), without the necessary changes in real capabilities will confuse rather than clarify public perceptions of U.S. policy at home and abroad, as was evidenced in the public reaction to Carter's Presidential Directive 59.* Eventually, the lack of results and the high costs of the incremental approach will force arms limitation talks back into a predominant position in U.S. security policy. The Reagan Administration will begin to sound more and more like its predecessor, and the mandate for strong defense and foreign policy will dissipate.
- The present inclination of the public and hence of Congress to support very high defense expenditures is a diminishing asset. The incremental approach does little to sustain public support for a strong national defense. Further, it creates the grave danger

* Presidential Directive 59 issued in 1980.

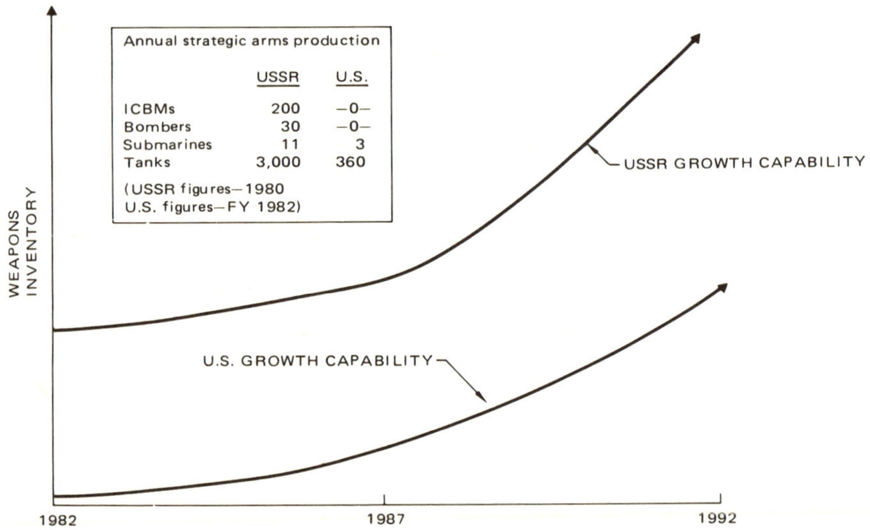

Figure 1. Incremental Approach

of a severe backlash against current proponents of increased defense expenditures if four years hence there is no perceptible, favorable change in the U.S.-Soviet military balance.

The bold approach offers strong possibilities for avoiding the pitfalls of the incremental option, while providing answers to the broad range of strategic problems that beset the U.S. and its allies.

- It replaces MAD with Assured Survival, so we can avoid leaving to future generations the sorry legacy of a perpetual "balance of terror" which in the end must favor the side most inclined to *use* terror.
- It moves the contest with the USSR from the arena where the Soviets have the advantages to one in which the U.S. has the advantages.
- It is a truly national, rather than a merely *military* strategy; it provides—in addition to security—promising solutions to economic problems (including the energy crunch), to the problems of development in the Third World, and to the erosion of spirit in the West.
- It can reverse the alienation of the public toward their military establishment by making the military effort understandable to the average U.S. citizen and compatible with his aspirations. At the same time, the new strategy will have the effect of disarming much of the antidefense/antitechnology sentiment in the Free World by offering a viable alternative to the continuous growth of destructive nuclear weapon inventories.
- It can accomplish all these things in less time, with less money, and with far more popular support than any available alternatives.

In our efforts to examine all dimensions of this new strategy we have considered international ramifications, including Soviet reactions and effects on existing treaties, macroeconomic effects, management options for implementation, cost implications, and the near term, nonspace collateral actions needed to support the new strategy. All of these factors are discussed in some detail in succeeding chapters of this study.

The results of the High Frontier Study, of course, cannot be as thorough as would be a similar effort by the government, but the results are already sufficiently definitive to support a U.S. strategy change.

CHAPTER II: THE MILITARY DIMENSION

The dimensions and severity of the military imbalance between the United States and the Soviet Union, as noted in the preceding chapter, need no further elaboration. Suffice it to say that Soviet success in their effort to add military domination of near Earth space to their already massive strategic power would insure many decades of Soviet global dominion. This prospect alone demands a vigorous effort on the part of the United States to maximize our current technological advantages to secure this new frontier for the defense of the Free World.

The imperative to meet the Soviet challenge in space is reinforced by the paucity of options available to us to meet the current threat by other means. In the recent debate over deployment modes for the MX missile, it became clear that a search for invulnerable offensive weapons is fruitless. While some modes are less vulnerable (e.g., submarines), none can be reliably predicted to remain invulnerable for more than a few years—especially when potential military systems *in space* are postulated as part of the Soviet arsenal.

Further we find that an attempt to redress the imbalances by incremental add-ons to land, sea, and air components of current U.S. forces requires very large and politically difficult expenditures. While perfectly reasonable arguments can be sustained for such expenditures (e.g., historical percentages of GNP devoted to defense) and perhaps should be persuasive, the fact is that a tailchase of the Soviets in production of current technology weaponry would probably be a strategic failure. The United States would start with a very low arms production rate from a seriously diminished arms production base, while the Soviets would proceed from a formidable arms production rate and base.

A TECHNOLOGICAL END-RUN

Any search for a "technological end-run" on Soviet military advantages leads inexorably to space. While there are promising technological innovations possible in current land, sea, and air weaponry, they are essentially product improvements unlikely to cause more than vernier changes in the overall strategic balance. It is in the area of space technology where the U.S. advantage can be decisive. While the Soviets lead us today in the *application* of space technology to military capabilities, we have a strong lead in *potential.* The Space Shuttle is the most obvious of our advantages, but the fundamental advantage lies in our ability to miniaturize and therefore achieve superior capabilities per pound of materials put into space.

Careful examination of U.S. military options in space leads to the firm conclusion that the priority effort must be to reestablish strategic defense. The only way that we can effectively shift the strategic framework of U.S. and allied thinking away from the Mutual Assured Destruction doctrine to Assured Survival is to deploy a global ballistic missile defense (GBMD). This can be accomplished effectively only in space. Partial withdrawal from the MAD doctrine is possible through deployment of point ABM defenses and renewed attention to civil defense. However, a true break with the all-offense, no-defense approach can be *decisive* only when it includes an effective spaceborne defense element. This is due to the fact that space offers the only potential for general, global strategic defense of the entire Free World at reasonable cost.

We are well aware that there have been cogent denials among military spokesmen that MAD has in fact been official U.S. doctrine, and some have attempted within the limits of their prerogatives to avoid the "mutuality" of Assured Destruction. The salient fact, however, is this: prior to the current Administration, the MAD doctrine has been sufficiently pervasive to prevent any serious attention to strategic defense options. All responses to increased strategic nuclear threats have been in terms of increased U.S. strategic nuclear offense.

We cannot, of course, allow our strategic offense to decay while we concentrate on strategic defenses. Political and military realities demand that we adopt balanced, strong, and mutually supportive offensive and defensive forces.

THE CASE FOR STRATEGIC DEFENSE

Certain articles of faith of the no-strategic-defense school of thought must be addressed and discarded. In the past, considerations of strategic defensive options have been attacked and defeated on these grounds:

- Defense systems are not useful in the nuclear era unless they are impermeable, i.e., *perfect*.
- Strategic defense systems are of little value if they can also be attacked along with the targets defended.
- Strategic defense systems are impossibly expensive.
- Strategic defense systems are destabilizing because they cause the opponent to believe we contemplate attack.

The notion that the strategic defense of populations and homelands is "bad" and "destabilizing" and that mutually retaliatory strategic offense forces are "good" and "stabilizing" is the "new wisdom" of the post-1945 period. Indeed, this particular complex of assumptions has been discernible in British and American strategic debate since the early 1900s, especially in the period just before World War II.

At that time a high level group—the Air Defense Research Committee—was established by the British government to look at the prospects for defending Britain from the German Luftwaffe. The question then was whether the Royal Air Force would allocate its funds to a "retaliatory" bomber force, in line with one school of analysis, or whether it would build up a homeland defense force of Spitfire or Hawker Hurricane interceptors, as well as radars and civil defenses. The threat of annihilation that the British government perceived at that time was parallel to what we see in Soviet nuclear ballistic missiles today. The British anticipated heavy Luftwaffe use not only of incendiaries but also of gas bombs against British cities. Some saw this threat as impossible to defend against with available or foreseeable technology. Fortunately for the Western democracies, the British government came down on the side of the interceptor force, the air raid wardens, and the radars that won the Battle of Britain. Unlike the American government of 30 years later, it did not embark on a wholesale ideological policy excursion in the direction of *mandatory* homeland vulnerability.

The rejection of the lessons of the Battle of Britain and the demise of serious U.S. consideration of strategic defense was signaled in a speech by Secretary of Defense Robert NcNamara on September 19, 1967. He specifically called for an impermeable defense or no defense at all:

> . . . it is important to understand that none of the (ABM) systems at the present or foreseeable state of the art would provide an impenetrable shield over the United States. *Were such a shield possible, we would certainly want it—and we would certainly build it. . . . If we could build and deploy a genuinely impenetrable shield over the United States, we would be willing to spend not $40 billion but any reasonable multiple of that amount that was necessary.* The money in itself is not the problem: the penetrability of

the proposed shield is the problem. (Emphasis added. Source: *Department of State Bulletin*, October 9, 1967.)

The original validity of the antidefense arguments can be readily challenged. Their validity is even more suspect in light of strategic defensive options available today.

With regard to impermeable or invulnerable defenses, there never has been nor ever will be a defensive system which could meet such criteria. Such perfectionist demands ignore the purposes of defenses and the effects of strategic defense on deterrence. Defenses throughout military history have been designed to make attack more difficult and more costly—not *impossible*. Defenses have often prevented attack by making its outcome uncertain. General Grant put a cavalry screen in front of his forces not because the cavalry was invulnerable to Confederate bullets or because he thought it could defeat General Lee, but because he did not want the battle to commence with an assault on his main forces or his headquarters.

It is this same military common sense that must prevail in our approach to strategic defenses today. Given the drastic consequences of a failed nuclear attack on an opponent, the critical military task is to keep a potential aggressor *uncertain of success* if not certain of failure. In the absence of defenses, the Soviet military planner has a rather straightforward arithmetic problem to solve to be quite sure of the results of a disarming strike against all locatable U.S. strategic weaponry—ICBM sites, airfields, and submarine bases. His problem is simply to ensure that he can deliver two warheads of current size and accuracy against each such target. If, on the other hand, the Soviet planner must consider the effects of a strategic defense, especially a spaceborne defense which destroys a portion of the attacking missiles in the early stages of their trajectories, he is faced with a problem full of uncertainties. He does not know how many warheads will arrive in the target area and—even more crucial—which *ones* will arrive over which targets. This changes the simple arithmetic problem into a complex calculus full of uncertainties; *such uncertainties are the essence of deterrence.*

Strategic defenses are imminently practicable and by no means impossibly expensive if the programs involved are not required to meet unrealistic standards of perfection or incredible postulated threats. A cursory review of combinations of spaceborne defenses, land based ABMs, and civil defense—while by no means *definitive* as to costs—indicates that a layered strategic defense system (see Figure 2) of decisive strategic importance can be devised which is relatively inexpensive when compared with some previously proposed offensive systems.

One attractive option open to us is to create a spaceborne ballistic missile defense quickly, using essentially off-the-shelf technology. We provide as an example of such a system the Global Ballistic Missile Defense (GBMD) system described in detail in Appendix C. This system is a multiple, unmanned satellite system which employs nonnuclear kill vehicles (already largely developed) to strike Soviet ballistic missiles in the early stages of their trajectories. It is a relatively rugged and uncomplicated system which can readily accommodate to improvement or enhancement which might prove technically feasible and attractive in the future.

A key issue which arises with regard to all space weaponry or, for that matter, space hardware in general, is survivability. Space vehicles are nearly invulnerable to some of the threats with which terrestrial systems must cope. These are most notably military attacks by troops, terrorists, and saboteurs armed with a wide variety of available weapons. On the other hand, space vehicles or weapon systems will never be completely invulnerable to a variety of deliberate attacks by a technically advanced adversary. In fact, space vehicles are by their nature delicate pieces of

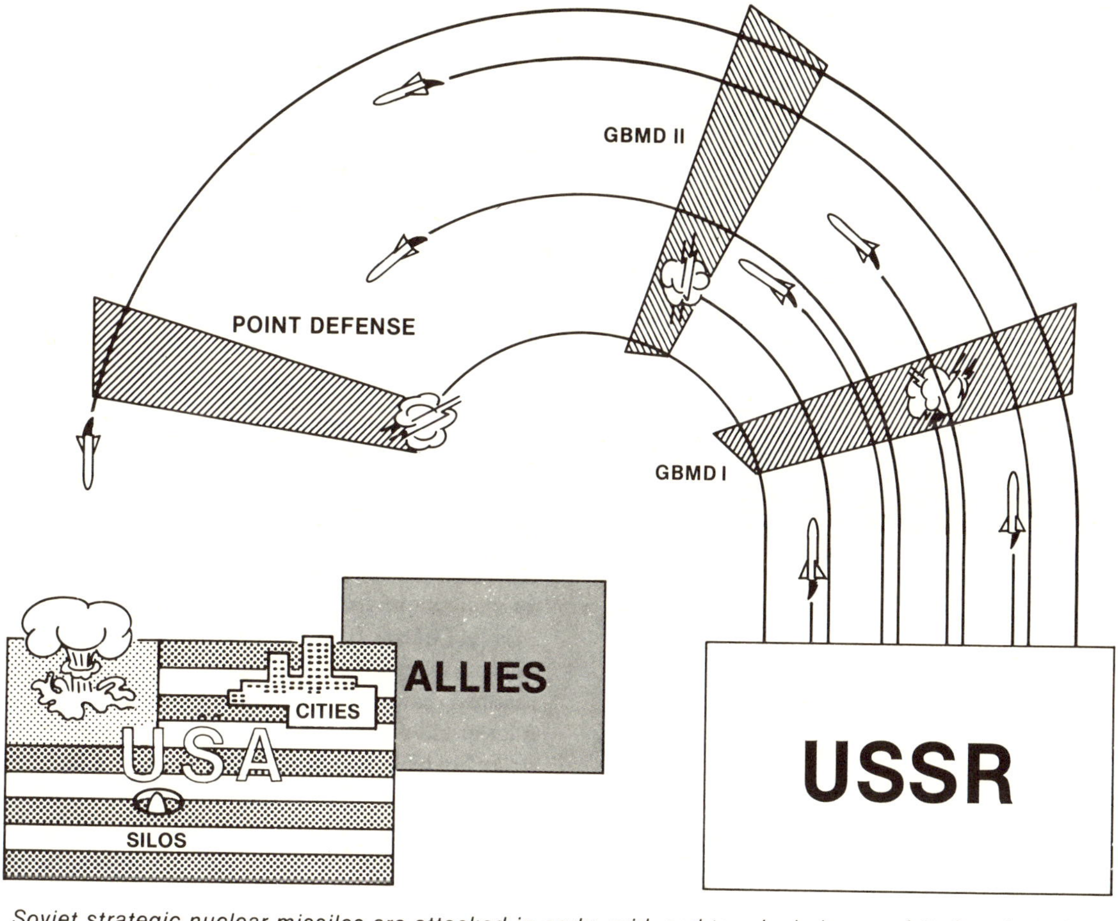

Soviet strategic nuclear missiles are attacked in early, mid, and terminal phases of their trajectories.

Figure 2. Depiction of Layered Defense

machinery which are presently exceptionally vulnerable to hostile action in the form of projectiles and nuclear explosions. In the future, they can become vulnerable to sophisticated beam weaponry, ground or space based.

Much of our current understanding of space vehicle vulnerability is based on the characteristics of our deployed individual, highly complex satellites, incapable of active self-defense. The concern about vulnerability is diminished when one considers a multiple satellite military system with each satellite capable of defending itself and many of its companions.

Other vulnerability concerns can be offset by technical means—primarily the hardening and maneuvering capability of the satellites. Clearly, the deployment of a defensive system in space would sharply reduce the vulnerability of operating U.S. space systems (communications satellites, intelligence gatherers, and Shuttle) as well as future military or civilian systems which we may decide to deploy. (For a fuller discussion of survivability, see Annex to this chapter.)

The first generation ballistic missile defense systems proposed for inclusion in a space based strategy are unmanned. However, we believe that the inclusion of military manned space vehicles is not only inevitable, but will be of great strategic value. The GBMD would be enhanced by the availability of a space utility vehicle providing man-in-the-loop, security, inspection, on-orbit repair, refurbishing, and adjustment. (One such vehicle, the high performance spaceplane, is discussed in Appendix D.)

The time required to achieve results is another key issue involved in the military dimension of this strategy. A key strategic point should be made at the outset. The enhancement of the deterrent to Soviet first strike against U.S. land based strategic systems begins with *partial* deployment of the GBMD, that is, long before the full system is operational. In essence, a definite decrease in Soviet ability to calculate the results of a first strike will occur when only a portion, perhaps 10 percent, of a spaceborne defense system is on station.

The time required to bring a spaceborne BMD to bear on the strategic balance depends heavily on the adoption of a management system capable of accelerating decision and procurement times which are now creating intolerable time lags in weapon system acquisition. (See Chapter VII, Implementation.)

Whatever the selection of spaceborne systems might be, we believe it also necessary to provide more immediate ground based ballistic missile defenses to hedge against delays beyond the critical period of the "window of vulnerability"* and to further complicate the problems faced by Soviet strike planners. The criteria for such point defenses should be that the system must be low cost and deployable in about two or three years. Its minimum essential function is to prevent a *confident* Soviet first strike against U.S. land based strategic missile silos. Such systems are available, and some examples are discussed in Chapter IV, Collateral Actions.

It is important to note that such simple ABMs have not been the focus of attention of past U.S. development. Much more sophisticated and expensive ground based systems would be required to provide an effective strategic defense in the absence of an effective spaceborne filter of attacking missiles. The low technology, low cost systems can provide the near term uncertainty of results in the Soviet planners' minds that is necessary to reestablish confidence in our deterrent. Their military value will increase sharply when confronted with only those warheads that leak through a spaceborne defense.

The military dimension of the strategy includes renewed attention to civil defense, which becomes a far more manageable problem in conjunction with active strategic defense. Finally, since the spaceborne defensive systems do not cancel out the need for an adequate balance of offensive systems, we have examined some options in that area, particularly cruise missile applications. These subjects are discussed in Chapter IV, Collateral Actions.

---

* A period in the 1980s when the Soviets could knock out the bulk of U.S. land based ICBMs in a first strike.

CHAPTER II ANNEX: SURVIVABILITY OF SPACE SYSTEMS

All space vehicles or installations now in operation or contemplated for the future are subject to attack by a determined enemy with an adequate technological base. This is true in any military endeavor. In fact, given the special vulnerabilities of space vehicles, if vulnerability to attack were the overriding consideration, there would not now be (nor would there *ever* be) any important space hardware in orbit.

Nothing is invulnerable to attack, space vehicles or otherwise, and the more complex the machine, the more vulnerable it is. Complex machinery on Earth has serious vulnerabilities not shared by systems in space. While space systems cannot be hidden from sight, encased in thick walls of steel and cement, or protected by barbed wire and soldiers, they are not subject to attack with ordinary weapons. It is highly unlikely that they could be disabled or rendered ineffective by natural events. In these regards, current space systems, although vulnerable in orbit, are more vulnerable in their Earth based links.

While relatively immune to certain Earth bound threats, current space systems have some special, serious, and thus far inadequately countered threats to their survivability. These threats derive from these characteristics of space systems:

1. They are physically fragile.
2. They are highly complex and delicate machines.
3. They cannot be hidden from the view of ground sensors.
4. They usually travel in fixed or nearly fixed orbits, making the exact locations highly predictable.
5. They travel at very high relative speeds which makes the impact with even a very small object highly destructive.
6. They operate in a vacuum which does not attenuate or scatter various forms of energy directed at them.
7. They are devoid of means for active defense.

Some technical actions have been taken by the United States to reduce these vulnerabilities. Electronic components have been protected against *far distant* sources of radiation (e.g., nuclear explosions), but they remain extremely vulnerable to less distant threats. Some vehicles are capable of changing orbits to avoid certain types of hostile action. Notwithstanding, our space systems remain essentially vulnerable, viable only as a result of Soviet tolerance which derives in large part from political considerations. In time of war, we must calculate that most of our currently undefended satellites would be destroyed in minutes, all of them within hours—a calamitous loss when one considers the extent to which we rely on our satellite systems.

Given the characteristics of current satellite systems, it is quite easy to list ways for the Soviets to attack them, ranging from throwing sand in their path with rockets to burning them out of space with exotic, futuristic beam weapons. We have noted a number of postulated attack modes and analyzed them individually. Some of the attack modes are possible for the Soviets to employ now. The USSR has a rather primitive but potentially effective antisatellite weapon system for single attack against a satellite in certain orbits. Further, they can now fire nuclear missiles at each of our satellites which would almost certainly destroy them. Within a few years, and within their *available* technology, the Soviets could probably develop capabilities to shoot down satellites over their own territory with direct ascent, nonnuclear missile attacks. Further in the future, beam weaponry—ground or space based—can pose even more serious threats to directly overhead targets.

Postulated attack modes, however, do not constitute real vulnerability problems for a given space system in *wartime* unless they cannot be mitigated or countered by technological or military measures. In *peacetime*, postulated attack modes do not translate into system vulnerability unless they can also pass the test of political credibility.

An examination of the survivability problem and its relationship to U.S. options for a spaceborne ballistic missile defense leads to these conclusions:

- Vulnerability of current U.S. space assets (intelligence and communications satellites and the Shuttle) sharply increases the imperative for an effective spaceborne defensive system which can defend itself, reduce the threat to other space systems, as well as defend ground targets against hostile objects transiting space, e.g. ICBMs.
- Defensive systems employing large numbers of less sophisticated satellites are far less vulnerable than those employing small numbers of more sophisticated satellites.
- An ability to provide mutual warning and protection among satellites in a ballistic missile defense is very important to survivability.
- The sooner a spaceborne ballistic missile defense system can be deployed, the better its survivability (long lead time systems are susceptible to long lead time Soviet countermeasures—real or postulated).
- Future U.S. deployment of more sophisticated beam weapon military satellites may be dependent for survivability on protection provided by a lower technology defensive system already deployed.

We have examined current and postulated future Soviet attack modes against U.S. space vehicles and find that they fall into two basic categories: peacetime attack and wartime attack. In wartime, the attack could be designed to destroy U.S. systems or impair their utility.

A much larger range of *current* Soviet capabilities could be brought to bear in peacetime. It is technically possible for the Soviets to attack U.S. satellites as they pass over certain areas of the Soviet Union with either direct ascent missiles or with their antisatellites (ASATs) armed with various nonnuclear destructive payloads. Some of these possibilities are difficult to offset technically without severe weight penalties to U.S. systems. Further, laser systems currently available to the Soviets could, after numerous attacks on each satellite, wear away protective shields and over time, destroy them.

However, these attack modes presuppose a Soviet willingness to risk the grave consequences (including war) of attacking U.S. *defensive* systems in peacetime. It is extremely doubtful that they would do so. Therefore, otherwise viable space defense options should not be eschewed on the basis of this class of threat.

The wartime threat is more critical. The most serious of *current* wartime threats to U.S. space systems are those of nuclear attack, either to destroy or to disable, using the effects of radiation resulting from nuclear explosions. Individual U.S. satellites or systems involving small numbers of satellites are more vulnerable to such attacks than are multiple satellite systems. There are technical shielding means available to partially guard against radiation effects—at least enough to require that individual nuclear attacks be mounted against each satellite if satellites are adequately dispersed in space. However, vulnerability to radiation cannot be completely overcome, especially if very large (100 megaton or larger) weapons were to be detonated in space in such a way as to maximize electromagnetic pulse (EMP) effects. The nuclear collateral damage to the Soviet Union especially through EMP makes this attack option highly unlikely.

If Soviet laser capabilities come to include very powerful pulsed systems, it is possible that they could destroy overhead satellites in a single attack. However, it is doubtful that such systems could, within the next decade or so, be able to destroy satellites coming over the horizon and thus shielded by much more of the Earth's atmosphere. This increases the survivability prospects for a multiple satellite defensive system.

There are important military responses to the wartime Soviet attack modes. All attack modes dependent for their execution on launch of a Soviet missile or satellite into space would be subject to attack by any space defense system which was also capable of performing its primary mission of attacking ballistic missiles. The Soviets could no longer count on mounting a successful attack against passive space systems. Even should the Soviets eventually create the means to attack a spaceborne defense system successfully in order to launch a subsequent strategic missile attack, all chances of destroying the U.S. deterrent on the ground would be lost. In these circumstances, launch on warning or launch under attack become both credible and feasible responses for the United States. The Soviets could not expect, after the attack in space, that the U.S. President would hesitate to respond to sensor warnings that a missile attack had been launched from the USSR. This fact alone would make a spaceborne defense of great strategic value.

CHAPTER III: NONMILITARY DIMENSION *

Space holds the prospect for a return to sanity in the national security sphere. It also holds the prospect for a revitalization of progress in the nonmilitary sphere. With a proper combination of space technologies, we can sharply improve the security of the U.S. and its Free World allies, and can also restore confidence in the ability of Free World economies to meet the challenges of the future.

Viewed from a historical perspective, few scientific and technical events have the stature of a genuine revolution in the course of human affairs as has the conquest of space. Man is still too close to his recent entry into the space age to comprehend fully the impact of this extraterrestrial expansion and the concurrent implications for economic growth, political influence, and national security. The conquest of space and its resources will be as significant in the economic expansion in the 21st century as improved ships, navigational techniques, and firearms were in the exploitation of terrestrial resources in the 19th century.

Space is the High Frontier that will be recognized as having the strongest influence on future strategies, both commercial and military. The success of these strategies will have profound effects on the resolution of many contemporary concerns, ranging from the availability of assured energy resources to meeting Third World economic aspirations.

In the past, government institutions and industrial organizations usually concentrated their planning and decisionmaking on the near term, most often less than five years. Space utilization strategies require long term planning, extending over the next several decades. Such long term commitments of resources imply a distant horizon for space programs. This may adversely affect the decisions regarding the development funding for the most promising option of the 21st century. For example, space missions such as the manned lunar landing or Skylab achieved their short term objectives but failed to provide any forward thrust for the next evolutionary steps. What is needed is the development of core technologies which can meet the requirements of a *multiplicity* of space applications, both near and long term, and can thereby maximize both the commercial and military uses of space.

A key factor in the success of future nonmilitary space programs may be international cooperation among Free World countries. This would require an international framework capable of coordinating, integrating, and managing the efforts of contributors in many parts of the world. With such cooperation, it is likely that the potential industrial uses of space can meet expectations without placing an undue burden on U.S. industry or taxpayers.

Industry, which must focus on the risk as well as the profit potential of ventures in space, will need the support of government. Industry cannot gamble with investors' funds. Therefore, when high risk, long term development periods or large capital requirements become necessary, joint industry-government cooperation will be essential.

In summary, the U.S. must be prepared to meet both the military challenges *and* the economic challenges of space. If successful, the U.S. will gain much more than the prestige of a Sputnik triumph or a Moon walk drama. The U.S. will garner the priceless advantage of security and the ability to supply valuable new services,

---

* For fuller treatment, see Annex.

manufactured products, and energy to the world.

In fact, the last-named opportunity—energy from space—has an especially promising potential. A decade of scientific studies and a recent comprehensive societal, economic, environmental, and technical assessment by the Department of Energy project that power could be beamed from space to Earth in large quantities. For example, it has been estimated that *one* solar power satellite (SPS) would be able to produce the electrical power output of up to *five* nuclear or coal plants. Further, the cost of this power would be competitive with terrestrial energy conversion options. If the development and implementation of solar power satellites were successful, electrical power could be supplied on a global scale. The economic gains and the independency from nonrenewable energy resources would be extremely valuable strategic assets.

It is time for the United States to embark on an astutely conceived commitment to exploit space. A piecemeal, on-again off-again approach, characteristic of past efforts, will not prove to be timely or economical. In contrast, Japan establishes long term goals and stays with them until they are successful. European consortiums are also increasingly setting long term goals. Where will we be, one or two decades hence, if we do less?

History is likely to recognize only two U.S. space efforts as crucial—our *national* commitment in the 1960s to be first on the Moon and our national commitment now to develop the strategic potential of space—if we seize the opportunity to do so. Security in space and access to space based energy, products and services, and an international realization of America's resumption of a leadership role in technology will be denied us, otherwise. This Administration is at the right place, and this is the right time to claim this place in history for the United States.

How do we exploit this opportunity? Initially we need to develop a more economical space transportation capability and to fund the low cost, preparatory and developmental stages of the most promising space industrial opportunities. Only after the means and economic viability of commercial ventures are established need large commitments of government or private funding be made.

The concurrent development of military and civil space support systems will produce valuable synergistic benefits to both civil and military programs because both depend on the same core technology. In the past the U.S. has erected artificial barriers between the two efforts. These barriers have been grossly detrimental to both space programs. However, under the present Administration, DOD and NASA are making renewed efforts to break down these barriers.

Another aspect of joint development which must not be overlooked is major investments in commercial space systems. It is not reasonable to expect individual nations, companies, or consortiums to create the space facilities necessary for industrial/commercial growth if such investments are not protected from hostile interference, attack, or seizure.

Our proposed space policy's most important, initial objective is the development of an improved Space Shuttle. Our first goal should be a substantial payload increase to garner the benefits associated with economy of scale, as well as a desirable increase in unit capability. Our second goal should be system cost reduction. The total achievable cost reduction using available technology is conservatively predicted by recent engineering studies to be ten to one (compared to the current Shuttle).

A recent technical assessment by NASA (I. Bekey and John E. Naugle, *Just Over the Horizon in Space, Astronautics & Aeronautics,* NASA Headquarters, May 1980) projects the following potential cost reduction: "The shuttle will not do better than $1000 to transport one kilogram to orbit, compared to only $5 to fly one kilogram in an airliner from Los Angeles to New York, although the energy requirements are the same. The cost of the equivalent electrical energy comes to only about 50 cents, leaving a lot of room for improvement." The NASA authors expect fully reusable vehicles and other systems to reduce the cost of space transportation "*by at least two orders of magnitude*." An optimized system would operate like an air cargo airline. It would operate on a rapid turnaround schedule assuring high utilization of ground support facilities and people.

The second immediate objective of our new space policy should be the creation of a manned space station in low Earth orbit. Such a station in near space would permit low cost, efficient development and testing of both civil and military system elements. The later deployment of a manned station at geosynchronous orbit is also assessed as a sound economic investment. Consequently, the first station may well be designed to serve as both an initial test facility and also as a "way station" for transitioning to a sustainable, manned presence in geosynchronous orbit. Eventually, the Moon itself may serve as a space station with important benefits for the construction of installations, such as SPS, in geosynchronous orbit.

Our third immediate objective should be the development of reliable, high capacity energy systems in space. The initial application of such systems would be the powering of other installations in space.

The final objective of our proposed space policy is the initiation of preparatory development of a selected number of promising commercial business opportunities. The term "preparatory development" was adopted to embrace the following considerations: the *ultimate goal* of all such effort is to establish independent and self-reliant businesses, any government efforts to "seed" these ventures should be oriented to this goal. The government's efforts should focus on preparing for the transition of these "seed" efforts to becoming *independently viable* commercial operations as soon as possible.

This approach should also suggest that system developments emphasize production, rather than research goals. The importance of this must not be underestimated. Research oriented programs seek to glean as much knowledge as possible from each development. However, undesirable effects usually occur, including increased development costs and program delays. In order to ensure that such adverse consequences are prevented, oversight boards which include businessmen should preside over any preparatory development destined to be a commercial business.

Means of meeting the first two objectives, improved space transportation and the construction of manned space stations, are being advanced by The Boeing Company, Martin Marietta Corporation, Rockwell International, and others.

Several proposals also exist for powering space installations: nuclear power plants in orbit; nuclear reactors on the Moon; and various options for the conversion of solar energy into electrical power. These systems require more study before specific recommendations can be made regarding preferred options. However, we can state that the strategic potential of energy from space is so enormous that vigorous research on these options is essential.

SPACE TRANSPORTATION AND SPACE STATIONS

Today's Space Shuttle, a returnable space transport, is proving to be more economical than any previous expendable rocket system. However, it was designed to use as much available hardware (like its strap-on boosters) as possible because its *development cost* had to be minimized at the time it was approved. This was due to NASA funding limitations. Consequently, its *economic* performance is less than technology would have permitted, even at the time it was designed. If we now develop a larger carrier and pay careful attention to ground cost minimization, engineering studies predict that future space transport costs could be cut to 10 percent or less of current Shuttle costs. The possibility of reducing Shuttle cost that much will profoundly reduce the costs (or increase the capability per dollar spent) of space defense systems. It will also greatly expand the range of business opportunities that will prove to be commercially viable because the costs are very sensitive to space transportation expense. Clearly, the development of a more economic Shuttle is the highest priority item for both the future military and the nonmilitary programs outlined.

A second generation Shuttle would most likely be a two stage vehicle with both stages fully reusable. Routine reuse would be patterned after cargo airline operations. Consequently, fast turnaround, high maintainability and minimum life cycle cost will be primary design requirements. First stage fuel would be a hydrocarbon like methane for economy. This stage could be of all aluminum construction—heat sink design rather than heat shield tiles.

The pacing development item would be the main booster engines. Based on system design studies performed for NASA, a 125-ton payload would require five or six engines of approximately Saturn F-1 size and type, but redesigned to provide longer life, low cost, and easy maintainability. A development cycle of only five years is possible with responsive program management. Total program cost would be in the vicinity of $12 billion.

The initial space station at low Earth orbit could be the space operations center (SOC) being advocated by NASA. This platform would be an operational base in space for the assembly and test of space equipment, repair of satellites, and the staging of equipment bound for higher orbits. Minimal scientific research would be conducted here.

With a 1982 go-ahead, a space operations center could be in place by 1986. Costs of approximately $6-8 billion are estimated.

SPACE BASED ENERGY SYSTEMS

Solar cell arrays in orbit may be used to collect energy from the Sun. This power can be relayed to Earth through microwave transmission. The microwave transmission system requires large antennas in orbit and on the Earth to be efficient. For example, in one design a 3000-foot-diameter transmitting antenna is part of the satellite and a four by five mile (elliptical) receiving antenna is required on the ground. If a high level of power, say 5,000 megawatts, is transmitted from orbit to Earth, then the total costs of the satellite power system, including the receiving antenna on Earth, can be apportioned over a very large amount of power.

Because of large scale operation of the system, delivered power costs are predicted to be competitive with coal or nuclear power plants. For example, if a $12.5 billion ($2,500 per kilowatt in 1981 dollars) system capable of 5,000 megawatt output were purchased, it might cost around $78 billion over 40 years to own and operate it ($12 billion in depreciation plus $21 billion interest at 12 percent, $33 billion earnings at 18 percent, plus $12 billion in operating expenses, taxes, and other costs). The station would deliver 1.6 trillion kilowatt hours of power over 40 years. Hence, the *average* cost of the power delivered is under five cents per kilowatt hour.

A comprehensive assessment of a representative space based energy system was conducted by the Department of Energy from 1977 to 1981. Their evaluation did not reveal any technological barriers. Continuation of system definition, refinement of initial socio-political-environmental assessments, and test of key system elements is the next step. Since cost attainment is vital, major emphasis in the program would be placed on cost control. Finally, *demonstration* of cost attainability for key system elements would be required *prior* to seeking funds for full scale implementation.

A conventionally paced research program would require spending about $30 million per year for the next three to five years. At the end of this period, commitment to pilot production of key items (to demonstrate *cost achievability*) and a limited space demonstration of promising technologies would be sought. The first full scale system would be built after 1995 if the system's promise is achieved.

An Administration decision to support a joint commercial/military space program has many political ramifications. It is believed that the majority are positive.

The dominant benefit to the U.S. of embarking on a joint military/commercial space development plan is that this country will then have a comprehensive space policy with well integrated *long term* objectives.

If the U.S. offers participation in these space efforts to our allies, many international benefits may accrue. For example, the proposed space defense is capable of protecting Europe and Japan and our allies would share in developing space based industry, so they would be likely to share in the costs of the program.

CHAPTER III ANNEX: SPACE INDUSTRIALIZATION

INTRODUCTION

Viewed from an historical perspective, few scientific and technical events have the stature of a genuine revolution in human affairs. The great watersheds leading to physical transformation and the way men have lived, thought, and acted, from the discovery of fire to the taming of the atom, must now include the conquest of space. Man is still too close to his entry into the space age to comprehend the potential impact of this capability to extend his evolution into space—a dimension that has until recently been unattainable, forbidding, elusive, and tantalizing in its unrevealed and unknowable promises. The extraterrestrial expansion of the human species will affect everyone. Already, the effects on communications and Earth observations are global. Success or failure in grasping the beckoning opportunities of space utilization will have as much influence on the destinies of nation states as the industrial revolution had on the development of the world as we know it today.

During the last 150 years, the industrial revolution affected lifestyles in every part of the world. New industries profoundly changed the relationships between nations. The 19th century empires disintegrated as the new political and economic dominance of individual nations, largely based on their technological progress, led to the East-West struggle for global influence and the emerging North-South dialogue.

Technological progress will continue to be the key to economic growth, political influence, industrial expansion, and national security. Thus the conquest of space and exploitation of its inexhaustible resources and unique characteristics will be as significant in supporting the economic and technological networks and determining international, political, and commercial relationships in the future as superiority in ships, navigational techniques, and firearms was in the development and exploitation of terrestrial resources in the 19th century. If we abdicate our lead in space, the result could be the rise once again of imperialism, using modern technology to the same ends as in the last century.

Some view space utilization as a diversion of funds from more worthwhile societal purposes, as an endeavor primarily of scientific interest, or as a form of entertainment covered by the mass media. Others view space as the High Frontier which deserves to be recognized as the single most pervasive influence on all future strategies, whether military or commercial. The success of these strategies will have most profound effects on the resolution of contemporary concerns, ranging from the availability of assured energy resources to meeting Third World economic aspirations.

In the past, government institutions and industrial organizations usually have concentrated their planning and decisionmaking on the near term, considering five to ten years to be long term. But space utilization strategies will have to be based on the projected consequences, i.e., scenarios of various program options extending through the next 50 years. Such long term implications of the use of resources beyond the Earth's biosphere for human benefit tend to give a futuristic implication to space programs. This connotation may adversely affect decisions regarding the funding for research and development required to exercise the most promising options for the 21st century. However, such options are required to help focus the near term space programs so that they provide information on which to base the next phase. One must avoid, as much as possible, space missions such as the manned lunar landing or Skylab, which achieved their objectives but foreclosed any forward thrust for the next evolutionary steps. What is needed is the development of *core technologies* which can meet the requirements of a multiplicity of space applications and which will lead to step by step advances in the commercial and military uses of space (Figure 3).

As important as the role of technology is in the planning and execution of specific space programs, economic, environmental, and societal issues must be considered in parallel to ensure the continuance of broad support for the space programs.

Industry, which must focus on the near term profit potential of ventures to be conducted on Earth or in space, will need the stimulation of government support. Without such support it will

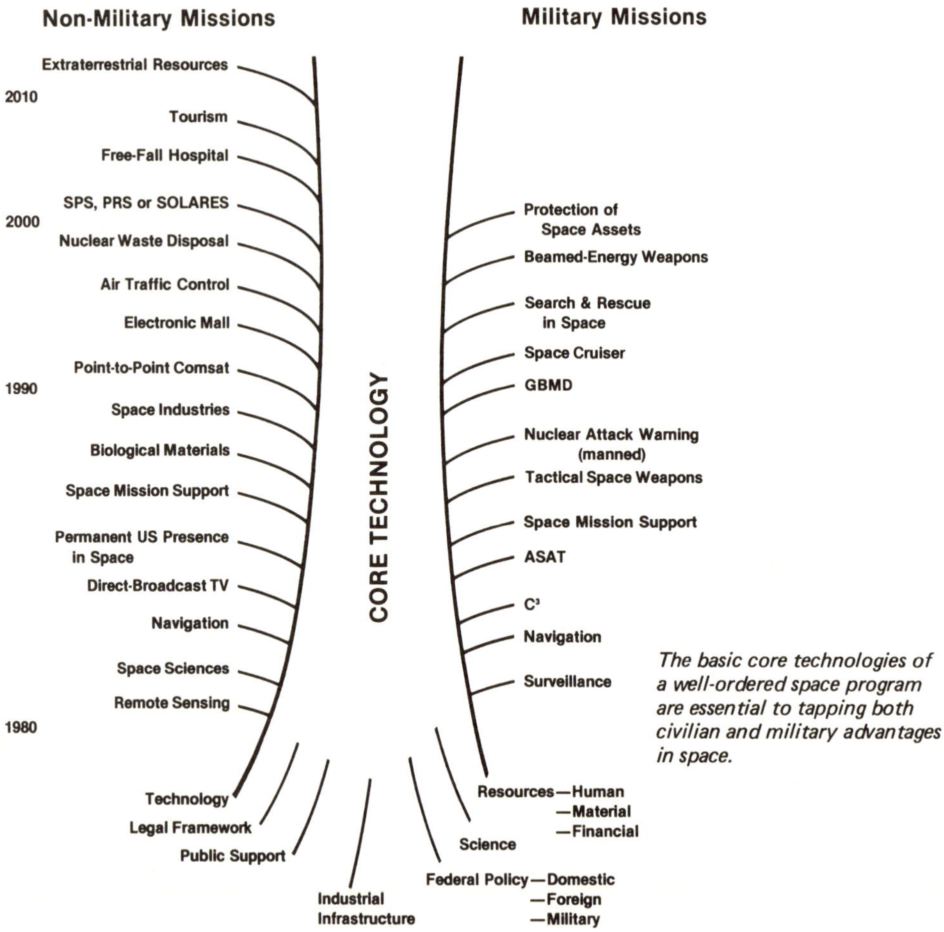

Figure 3. Core Technology and Space Applications

not embark on and participate in specific activities requiring long term, high risk investments. Joint industry-government cooperation and planning will be essential to achieve the longer term nonmilitary space program goals of creating profitable markets for industry.

Inevitably, in today's economic climate, space activities are going to be assessed more critically than they were two decades ago. But financial constraints should not be the sole determinant to the planning of industrial activities in space. A close coupling between these activities and national security implications is imperative. Beyond that, the activities should be part of the strategy to find synergistic solutions to humanity's most pressing problems pertaining to energy, the environment, and resources.

What is necessary today is to gain a broad perspective of the potential of space and to engage in strategic planning, guided by Robert Goddard's statement: "It is difficult to say what is impossible, for the dream of yesterday is the hope of today and the reality of tomorrow."

INTEGRATED SPACE MISSIONS

There is a wide divergence of views regarding the long term global impacts of space missions; the influence of advances in science and technology on future commercial, industrial, and military activities; their competitiveness with similar activities performed on Earth; and the scale, timing, and effectiveness of investments in space programs by industrialized nations in response to idealistic visions, pragmatic considerations, and political realities.

The U.S. space program has pursued two separate paths. One is representative of the nonmilitary activities carried out by NASA and more recently, by industry, primarily in communications. The other is the Department of Defense (DOD) military activities which have been recognized as being critical to U.S. national security.

The nonmilitary space missions fall into the broad categories of information, energy, industry, services, and science. The military space missions are concerned with intelligence, defense, and offense. Although the nonmilitary and military space program budgets are about the same, there have been only limited interactions between the two programs. It is obvious that there is considerable synergism between these space missions. This was recognized in the joint NASA/DOD development of the Space Shuttle and deserves increasing attention as more ambitious missions which could use a similar technological base are undertaken.

In view of the significant investments required for future manned and unmanned space missions to achieve military and nonmilitary goals, an integrated long term strategy based on incremental advances of technologies of increasing scope and effectiveness is essential. Such a strategy is being pursued by the USSR, where it is at times difficult to discern the boundaries between military and nonmilitary space missions, because the development of specific technologies is common to both.

The U.S. nonmilitary space program has been characterized by well publicized missions (e.g., Apollo, Skylab, and Voyager) which were undertaken to meet specific political, technological, or scientific goals. Despite the wealth of information obtained, these missions resulted in a dead end because an integrated long term strategy for space applications had not been evolved.

Space missions can be viewed as branches of a tree trunk represented by core technologies as the strong roots required for future growth (Figure 3). The core technologies should be developed to support nonmilitary and military missions, with each branch providing increased capabilities and requiring more substantial capital investments. The success of near term space missions would reduce the risk of introducing subsequent missions and would also establish the requirements for the development of advanced core technologies to ensure that research and subsequent development will be supportive of the largest number of future mission goals. For example, the goal of achieving a permanent U.S. presence in space would have significant nonmilitary and military mission implications. Military space planning today does not include the construction of large space structures, advanced robotics, power plants in space, or manned operations in geosynchronous orbit. The commonality of core technologies serving both nonmilitary and military mission goals should be analyzed so that the space program budgets can be most effectively used in support of a national space policy. The evolution of a national space policy which recognizes the potential of space to meet a variety of national goals deserves a high priority.

SPACE RESOURCES

The projected benefits of nonmilitary missions are closely coupled to the availability of the unique resources of space. Space is not just a medium to pass through while observing and supporting activities on the Earth's surface. As important as it may be to utilize the inherent characteristics of space, its resources are key to a broadening range of applications.

The primary space resources include the following: absence of gravity in satellites in free fall in various orbits; gravity forces, which determine the shape of orbits of satellites (for example, geosynchronous orbit, which has the unique characteristic that an object in orbit is stationary with respect to any desired location on Earth); high vacuum, which cannot be easily produced on Earth; an infinite heat sink of only a few degrees above absolute zero, permitting heat to be radiated from a satellite to deep space; electromagnetic radiation, primarily solar energy; presence of magnetic and electrical fields; an unobstructed view of Earth from low Earth, Sun-synchronous, elliptical, and geosynchronous orbits; orbits around the Moon, Sun, and other planets; and the availability of extraterrestrial materials from the Moon, asteroids, and other planets, which could forever remove the limits to growth of human civilization on Earth.

The wide range of nonmilitary space missions which are being performed already or are in various planning stages is shown in Figure 4. The broad mission categories, including information acquisition and dissemination, energy supplies, industrial activities, commercial services, and scientific investigations, can be classified according to orbits, e.g., low Earth and geosynchronous, lunar location, or missions which encompass the solar system. The striking feature of current nonmilitary space missions is their global impact and the growing level of investments in information services and communications. These missions are precursors of future industrial and commercial activities which can create profitable opportunities for private investment and lay the foundation for a dramatic expansion of space activities.

The industrial uses of space, in addition to information services and communications, include the conversion of energy sources in orbits, the beaming of power from space for use on Earth, processes to produce unique materials which take advantage of the characteristic of the space environment, disposal of hazardous materials by selecting orbit trajectories which would result in collision with the Sun, space construction in support of industrial activities, the exploitation of mineral resources on the Moon and in the asteroids, and people oriented services.

Figure 5 lists the space industrialization activities which could take advantage of space resources and indicates the major technical hurdles which will have to be overcome to achieve the goals related to information, energy, materials, and

Location	Information	Energy	Industry	Services	Science
LEO	• Remote sensing • Mineral resources • Agriculture • Fisheries • Search and rescue • Disaster relief • Earthquake prediction • Hydrology • Fire detection • Pollution monitoring • Border surveillance • Cartography	• Sun-synchronous SPS • SOLARES	• Biological materials • Solid-state devices • Advanced alloys • Improved magnets • Optical components • Superconductors • Freefall casting • Space construction	• Space mission support • Tourism • Freefall hospital	• Optical astronomy • Infrared and ultraviolet astronomy • X-ray astronomy • Materials science • Biology • Solar physics • Geophysics • Oceanography
GEO	• Communications • Direct-broadcast TV • Point-to-point communications • Electronic mail • Education • Weather forecasting • Navigation • Search and rescue	• Photovoltaic SPS • Thermal conversion SPS • Fast breeder reactor • Power satellite • Power relay satellite	• Hazardous materials • Space construction	• Space mission support • Air traffic control • Night illumination • Biological isolation	• Radio astronomy • Geophysics • Solar wind studies • Meteorology
Moon		• Solar power breeder reactor	• Mineral resources		• Radio astronomy—(farside) • Planetology • Geology
Solar system		• Nuclear waste disposal	• Asteroid resources		• Planetology • Geology

Figure 4. Nonmilitary Space Missions

Activity	Information	Energy	Materials	People
Major space advantage	• View Access	• Solar flux	• Low g • High vacuum	• Uniqueness
Major technical	• Antenna size—10m to 100m • Power — 21 to 10,000 kW • Data processing	• Power output—1 to 5 GW • System mass — 10^4 tons • Cost—$\$10^{10}$ • Transport cost — < \$20/lb LEO • Environmental effects • Societal issues	• Proof of theory • Production and process development • Power — 10 to 10,000 kW • Transport cost — < \$100/lb LEO	• Transport cost ⩽\$25/lb • Habitation
Timing for significant revenues	• Present > \$300 M/yr	• 2010 +	1990 +	• 2010 +

Figure 5. Space Industrialization Activities

people. For example, enhanced data processing and information exchange capabilities would require increased antenna sizes and power supplies. Utilization of solar energy could lead to the development of solar power satellites to beam power to other satellites in orbit and to supply power to Earth to supplement terrestrial power generation. Large mirrors could reflect solar radiation to desired areas on Earth so that terrestrial solar energy conversion facilities could be utilized 24 hours a day. Unique materials could be produced in the near zero gravity and/or high vacuum in orbit for use in space construction projects, or on Earth. Lunar or asteroidal materials could supply material requirements of space construction projects.

GROWTH POTENTIAL

Efforts are being directed to confirm the possibility of producing unique materials in space. Once it has been demonstrated that such materials can be produced in space, and information about their structure and properties has become available, production methods and processes could be developed and space industries established. The low gravity conditions achievable in orbit, possible attractive features of living in space, and utilitarian motives associated with space industrial activities may result in a gradual increase in human habitations in space. Although it is too early to project whether large scale human migration to space will take place at some future time, demonstration of habitability of space without adverse effects can be expected to result in a permanent human presence.

The timeframe for the growth of space industrial activities extends well into the 21st century, with significant revenues projected for each one of these activities. The revenue potential will, to a large measure, depend on the successful development of economic and reliable space transportation systems. The thrust of space transportation system development is exemplified by the Space Shuttle, whose goal was to achieve a significant reduction in transportation costs to low Earth orbit. There is no inherent technical barrier to the development of an economic space transportation system which could approach airline-type operational procedures in support of space industrialization.

Space systems today extend to only a few meters in size, but systems measuring several thousands of meters are already being analyzed. The largest satellites in orbit have masses of a few tons but construction of systems with masses of several thousands of tons are being projected. Manned activities in space have been carried out by highly skilled and trained astronauts staying in space for a few months, at most. Semipermanent occupancy of space by thousands of workers would be required to achieve the goals of expanded space industrialization activities.

In a little more than two decades since the dawn of the space era, man has penetrated outer space, landed on the Moon, orbited hundreds of satellites, and obtained valuable and beneficial global information. It is inconceivable that this evolution of space activities will not grow exponentially as space industrialization opens new markets, demonstrates expanding opportunities for business ventures, and becomes the arena for national cooperation as well as competition. It is clear that numerous political, social, legal, and financial challenges will have to be met both on a national and international scale so that the tangible returns will be of widest benefit to society and reward to those participating in the creation of a new industry.

Although projections of markets and potential revenues are based on assumptions regarding the future course of the development of space industrialization, the magnitude of markets for information and energy related activities is sufficiently large (Figure 6) that even if annual revenue fails to increase as rapidly as projected, the revenues of

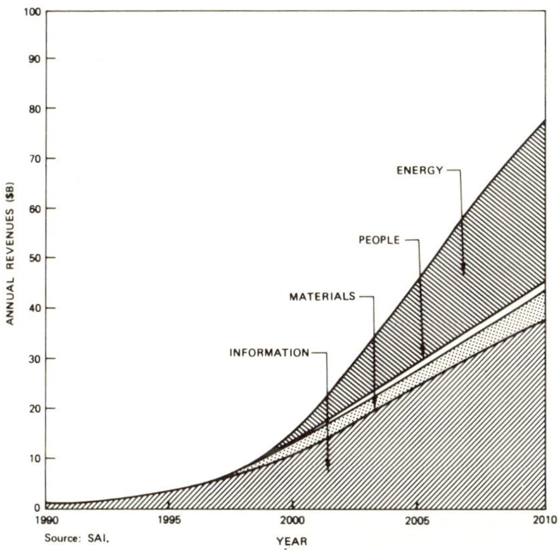

Figure 6. Projected Space Industries Revenue Comparison

space industries could be among the fastest growing of all industrial activities of the 21st century. Space industrialization could have as profound an effect on nations in the 21st century as the industrial revolution had on the political realignment of nations in the 19th century.

INTERNATIONAL PARTICIPANTS

The growth of space industrialization capabilities is not restricted to only a few nations, but includes an increasing number of participants (Figure 7).

The Soviet Union has demonstrated its commitment to a vigorous long range expansion of its space capabilities with emphasis on manned space operations lasting many months. These are the precursors of space industrialization activities, which parallel continuing commitments to substantial military operations in space. Several European countries are increasing their space related activities and Third World countries are increasingly recognizing their stake in the successful outcome of space industrialization endeavors. One of the striking examples of a growing commitment to space programs is the Japanese space industry's projection that annual sales will grow from one hundred billion yen per year in 1980 to one trillion yen in the mid-1990s. The Japanese expect to share not only in the growth and demand for various satellites, but also in the manufacturing of various products in space and in the development and launching of rockets. The Japanese realize the value of a high technology program as a stimulus for their economy and as a means to improve the quality of life in Japan. The Japanese report states: "The world is in a major transitional stage to enter the space utilization and Japan, too, should ride on the wave of the future and move steadily forward—it is necessary to plan for Japan's development of the space industry in line with the U.S. and Europe from the standpoint of a long range view and global outlook."

Another example of international interest in future space activities is the studies of the solar power satellite concept which are being carried out in parallel with those in the United States and in Canada, England, France, Germany, Japan, and the Soviet Union, as was reported at international meetings of professional societies. In addition, Austria and India have established industry working groups under government auspices which follow the progress in this field.

Already a number of organizations whose aim is to advance space industrialization have been founded (Figure 8). The largest and most successful international organization supported by governments is INTELSAT. It is most likely that international, governmental, and commercial organizations aimed at space industrialization will proliferate as transportation costs decrease, orbital operational capabilities increase, and supporting technologies are developed. Among the most important technology developments is advanced robotics, which may make it possible to reduce the requirement for labor-intensive activities in space

	General				Unmanned				Manned				Unmanned			Manned			
Nation / Capability	Data handling	Ground station(s)	Launch facility(s)	Suborbital launch	LEO launch	LEO return	GSO launch	Orbital rendezvous	LEO launch	LEO occupancy	Orbital rendezvous	Orbital propellant transport	Earth observation	Communications	Navigation	Test and experiment	Maintenance and repair	Remote control	Materials processing
United States	X	X	X	X	X	X	X	X	X	X	X		X	X	X	X	X	X	X
USSR	X	X	X	X	X	X	X	⊗	⊗	⊗	⊗	⊗	X	X	X	⊗	⊗	X	X
China (PR)	X	X	X	X	X	X							X	X					
France	X	X	X	X	X	?							X	X					X
India	X	X	X	X	(X)								X	X					
Japan	X	X	X	X	X		X	X						X					
ESA	X	X	X	X	X		(X)			(X)			X	X		(X)			X
Other	X	X	X	X	X								X	X				X	X
Total number of nations	111	39	24	15	9	3	4	2	2	3	2	1	7	13	2	3	2	3	5

(X) To be demonstrated.
X Capability has been demonstrated.
⊗ Currently unique capability.
Source: SAI.

Figure 7. Global Space Industrialization Capabilities

		Formation date	Organization	Composition	Activities
Governmental (international)		1964	• Intelsat	104 nations	• Intelsat series
		1971	• Intersputnik	9 nations	• Molniya
		1975	• European Space Agency (ESA)	10 nations	• Many (including Ariane as shuttle competitor)
		1975	• Nordic Telescope Satellite Committee (NTSC)	3 nations	• Nordsat (Ariane-launched)
		1976	• Arab Satellite Telescope Organization (ASTO)	20 nations	• Arabsat
		?	• International Maritime Satellite System (INMAR SAT)	40 nations	• Not yet ratified
Commercial	Multinational	1961	• Eurospace	96 companies; 31 banks	• Promotes Eurospace independence (especially launch vehicle)
		1971	• Cons. Ind. Fr-Al. Symphonie (CIFAS)	4 French companies 4 FRG companies	• Symphonie series
		1976	• MESH	5 European companies	• OTS/ECS
		1977	• Eurosatellite	3 European companies	• H-SAT
		1977	• ARCOMSAT	2 French companies 1 FRG company	• Proposing Arabsat
	National	1975	• Orbital Transport and Rocket Co. (OTRAG)	FRG inv.	• Private launch vehicle (Libya)
		1976	• Spacelab Utilization Working Group (ANS)	3 German companies	• Spacelab exploitation

Source: SAI.

Figure 8. Organizations Aimed at Space Industrialization

and place increasing reliance on automated assembly, construction, and operations of space industrial facilities.

STRATEGIES FOR SPACE INDUSTRIALIZATION

The broad goals of space industrialization are to provide options for the most vexing challenges found on Earth, the dwindling natural resources, environmental degradation, and the aspirations of humanity to achieve a higher standard of living. Space industrialization is synergistic with national security because extensive space operations will have a strong influence not only on the leadership of U.S. industry in high technology, but will also provide the industrial infrastructure required for both nonmilitary and military space operations. The construction of large information systems, plants for processing materials in space, and large energy conversion systems for use in space or supplying power to Earth are unlikely to be achieved by an Apollo-type national effort. Rather, they will be achieved by a vigorous national space program which recognizes that a series of well-defined evolutionary steps, solidly founded on the development of core supporting technologies, will be required (Figure 9).

The year 2000 has been chosen as an arbitrary dividing line for those space applications and core technologies. There is a reasonable consensus that they could be developed during the next 20 years. The investments indicated in Figure 10 assume an average funding level of about $5 billion per year. The revenues during the first 20 years of nonmilitary space applications may achieve only 20 to 50 percent of investments but set the stage for the more extensive applications indicated in

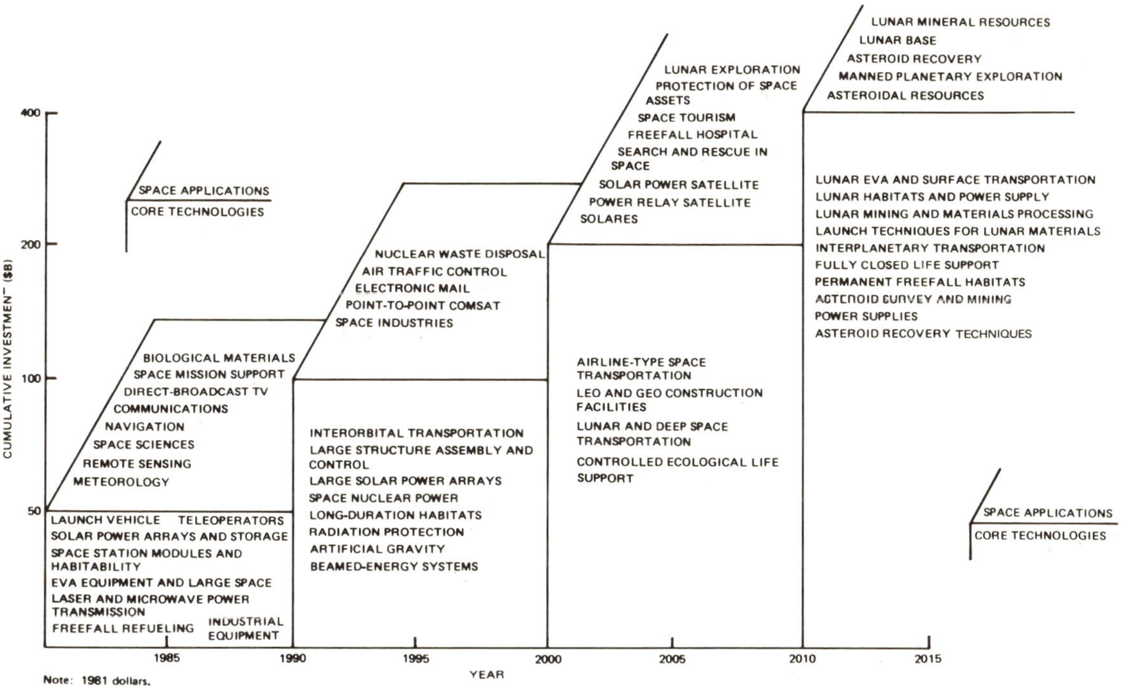

Figure 9. Nonmilitary Space Application

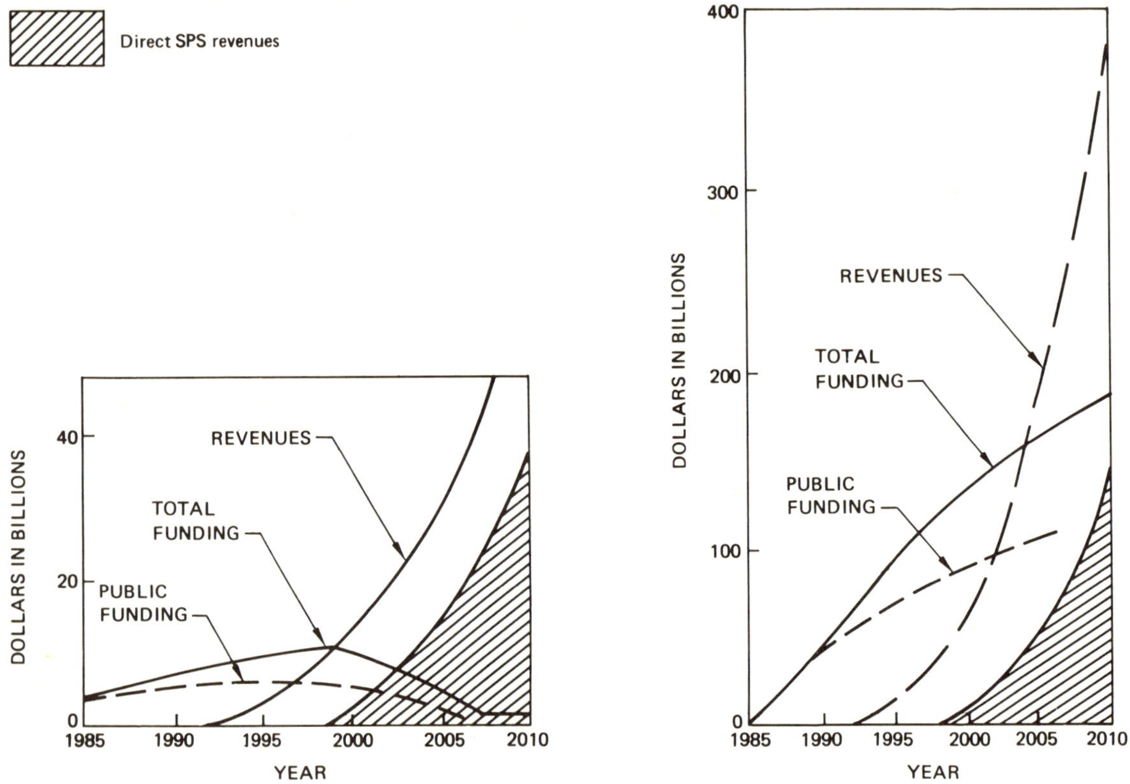

Annual Revenues and Expenditures **Cumulative Revenues and Expenditures**

Figure 10. Space Industrialization Program Scenario—Cost and Revenues

Figure 9, when revenues can be expected to exceed investments.

Space industrialization on a greatly expanded scale could be expected to occur after 2010, when extraterrestrial materials may bring about a new era in human activities in space. This may provide solutions to the challenges concerning the availability of energy supplies, irreversible environmental effects, and the depletion of natural resources.

Space industrialization implicitly assumes that industry will initiate the practical application of research results and technological developments which have been publicly funded. The challenge to a national space policy will be to reinforce and stimulate the growth of the market for the products and services associated with space industrialization, rather than substitute government commercialization activities. Should government activities be carried too far or any agency remain involved too long beyond the research and development stage, private investments may be discouraged and the leading edge of space industrialization blunted.

Although the emphasis on economic justification for specific space industrialization thrusts is appropriate, environmental and societal issues must also be considered. The social costs of environmental impacts, whether occurring on the Earth, in airspace, or in orbit, will have to be established so that the benefits of the systems to be used in a specific space application can be weighed against potential dangers to human health, resource commitments, potential destruction of valued natural resources, and intangible effects which may influence the quality of life.

The benefits and costs of specific space industrialization programs are not likely to be uniformly distributed. It is more likely that they will be concentrated in certain segments of society and the economies of industrialized countries. Individuals, corporations, institutions, and sectors of industry will react to the costs and benefits of specific developments as they perceive them. As a result of these perceptions, political pressures may arise which would have a pronounced effect on the direction of a specific development program, its schedule, and its ultimate success, in terms of both public acceptance and return on the investment.

Expansion of space industrialization activities can be expected to heighten public concerns on issues such as centralization and decentralization of control of a specific project and public involvement in project review. To alleviate public concerns and enhance public acceptability, information on goals, costs, and benefits of space industrialization activities will be desirable, particularly if international participation is to contribute importantly to the success of space industrialization. Unlike Project Apollo, space industrialization focuses on a multiplicity of objectives. Each one is to be achieved in a specified timeframe, requiring predictable investments and resulting in concrete benefits.

The "one small step for man, one giant leap for mankind" taken in July 1969 is not the appropriate analogy for many of the activities which will have to be integrated over extended periods to achieve the evolutionary advances which space industrialization implies. A better analogy may be the stepwise advances in technology achieved during the industrial revolution which continued during most of the 19th century, and which laid the foundation for the advances of technology in the 20th century, which in turn have to serve the needs of the global civilization in the 21st century.

In summary, the strategies for space industrialization include the following:

- Adoption of a long range view and global outlook.
- Integration with national space policy planning.
- Consideration of national and international market forces.
- Cooperation between industry and government to achieve jointly agreed upon goals.
- Evolution of institutional structures which acknowledge the needs of the public to benefit from space industrialization activities, the legal and regulatory framework so that the space industrialization can achieve its planned benefits, and mechanisms which will allow both public and private investments.

A national space policy which recognizes the inescapable realities of the future of space applications is required so that the U.S. technical leadership in space can be maintained and translated into practical industrial terms to strengthen the U.S. economy, develop exportable products and services, create new jobs, spur the economy, and demonstrate that the national agenda for space will additionally serve national goals and be designed to be of global benefit so that the 21st century opens the promise of the inexhaustible resources of space.

CHAPTER IV: PART A GROUND BASED ANTIMISSILE DEFENSES

INTRODUCTION

During the past several years, many proposals have been made for rapidly deployable, simple, and inexpensive ground based ballistic missile defense (BMD) systems. Most of these proposals have been studied by various elements of the Department of Defense. In general, objections to the proposals have been that they are complex and expensive. In any event, funding of the BMD program has not been adequate to permit experimentation; therefore, all of these proposed systems have been disposed of with paper studies.

More needs to be done.

THE ESSENTIAL PROBLEMS OF POINT DEFENSE

For decades, there has been a strong bureaucratic imperative which has worked against point defenses, i.e., systems that would defend a single U.S. ICBM silo. The best "cost effectiveness" case is made for BMD systems that defend the largest possible number of assets with each defensive unit deployed. This search for "cost effectiveness" drives the BMD developer in the direction of area defense, that is, defense of a sizeable piece of geography containing a large number of potential targets.

The more geography one attempts to defend, the more complex and expensive the antimissile system becomes since the system must be able to engage multiple numbers of warheads simultaneously and over a large area. The most difficult problem of all is that the farther away (and higher up) the incoming warheads are, the more difficult the problem of discriminating between real warheads and decoys. This problem has not been solved to date—hence all decoys must be engaged as well. These factors both drive up the complexity and costs of area defense systems and degrade system performance (kill probability) dramatically.

Worse yet, if an effective wide area BMD system could be developed, the BMD system itself becomes the most attractive and *vulnerable* target within the area defended. It is a lucrative target for the Soviets because its destruction is key to successful attack on the defended assets. It is also a tempting target for budget cutters not only because of large cost totals, but also because the entire complex system must be fielded before any military result can be produced.

The system is especially vulnerable to attack because of the softness of its critical radar components. They are subject to destruction by weapons with less accuracy and lower yields than those required to attack the defended assets. Given current Soviet submarine launched missile characteristics, a Minuteman missile complex need not be defended against them. But if that complex included an area defense system, it would have to be protected against *all* Soviet ballistic missiles (e.g., submarine launched missiles), raising once again the complexity and cost of the BMD system.

The point defense system has some significant advantages over area systems. The radars required are relatively simple and inexpensive and need cover only that small threat cone through which a warhead aimed at its single protected silo must come. It can operate almost autonomously (*automatically* when nonnuclear kill mechanisms are used).

And a point defense becomes militarily effective on a unit-by-unit basis. Policymakers could choose to defend any number of hardened assets (e.g., missile silos) in any geographical distribution. Further, in the more simple point defenses the problem of the vulnerability of radar systems is solvable and the entire defensive system can be emplaced within the already secured real estate occupied by current U.S. ICBM silo installations.

SUMMARY OF SYSTEMS CONSIDERED

Both nuclear and nonnuclear, quick fix systems have been proposed, including such nuclear system ideas as planting nuclear charges around ICBM fields for detonation at an appropriate time to neutralize incoming reentry vehicles (RVs) with clouds of dust and debris. Other proposed nuclear systems include interceptors with small nuclear warheads. There has been a strong, persistent antipathy to nuclear systems by DOD, Congress, and industry, on the basis that deployment would never receive popular support. Indeed, it would be difficult to muster political support for the deployment of any new types of nuclear weapons on U.S. soil. Further, the use of nuclear warheads greatly complicates command and control problems. The defense system could not react without Presidential authorization.

The nonnuclear, quick fix systems have generally incorporated interceptors of two types: those guided after launch, and those unguided after launch. Guided interceptors are considered more effective because, in the high winds and shock waves of a nuclear environment, their course can be altered to obtain successive intercepts. The unguided interceptors could destroy one RV, but follow-on interceptors could not adhere to their predetermined courses in the highly turbulent atmosphere existing in the first few seconds following the first nuclear detonation. (These are generalized statements relating to carefully structured ICBM attacks.) The cost of an unguided interceptor system could be appreciably less than that of a guided interceptor system, and the time required for development and production could be less than that for the guided system.

The radars associated with nonnuclear, quick fix systems are generally postulated as relatively low power (10 to 20 kilowatt average power, 100 kilowatt peak power), with other design requirements well within current technology.

Typical required characteristics of quick fix systems are that they can be sufficiently hardened against the blast and radiation generated by a one-megaton burst at about a 5,000 foot altitude and that they can be essentially unmanned and automatic in operation.

The following systems offer the most promise of a quick fix. Details are available from proponent firms or government agencies.

LOW ALTITUDE DEFENSE SYSTEM (LOADS)

This area defense system is funded in the defense budget and development has been under way for several years. Principal contractors are Raytheon (radar), TRW (software), Martin Orlando (interceptor), and McDonnell-Douglas (system integration). The system incorporates radars of modest power and guided interceptors with nuclear warheads. Later it may incorporate a nonnuclear warhead, but only a token effort is under way on this feature. LOADS is included here because it has been funded, albeit modestly; early development efforts have already been accomplished. Test flights with hardware (except the nuclear warhead) may be possible somewhat sooner than the other systems described here.

LIMITED AREA ABM SYSTEM

This system is under study by Vought Corporation. It incorporates a phased array Patriot radar variant and guided nonnuclear interceptors with flechette warheads. The interceptor is a derivative of the Vought T-22 (Lance follow-on) missile having a 1,050 pound payload and Phoenix-type seeker. Designed to defend cities and high value targets, the system could also defend Minuteman and MX.

SWARMJET

A concept under study by Tracor MBA, SWARMJET incorporates a radar system (using range-only radars), deployed in a trilateration scheme forward of the defended area; rapid fire launchers with several hundred projectiles per launcher; and projectiles that are ballistic rockets of high velocity (about 5,000 feet per second), any one of which can achieve a kinetic energy kill on an RV. The system has been under study for the defense of silo based Minuteman (see Annex B for details).

SANDIA SYSTEM

This concept was developed by Sandia National Laboratories at Albuquerque. It employs range-only radars in a trilateration scheme and unguided nonnuclear rockets which are directed to a point in space where the warhead is detonated at the predicted RV location. Application is to the defense of silo based Minuteman or MX.

DISCUSSION

All of these systems achieve low altitude intercept, except for the Limited-Area ABM System which proposes an intercept altitude of 50,000 to 75,000 feet. Only LOADS has a nuclear warhead, and all systems are hardened against blast and radiation.

Although the requirement for ballistic missile defense is becoming more widely recognized, all of these quick fix systems are today little more than studies. The problem now is not whether these paper systems would be effective, but whether hardware concepts (missiles, radars, launchers, C^3) are feasible in the areas of performance, producibility, and affordability.

To address this problem, greatly increased experimentation is needed. The R&D budget allocated to such activity should be greatly increased over the next two years, the amount being largely determined by the schedule on which prototype missiles and launchers can be produced. Radars should be as nearly off-the-shelf as possible.

CHAPTER IV: PART B FRAGMENTED ORGANIZATION IN U.S. SPACE PROGRAMS*

The fragmentation of the American military space program is considerable by any measure. A brief look at who does what for whom in space, from the operational level up to the policy or planning level, exposes the complexity and organizational overlap of space activity.

Of the three services, the Air Force operates with the greatest number of separate offices for space activity. At least four major Air Force commands are involved in space operations.

STRATEGIC AIR COMMAND (SAC)

This command manages and operates early warning and surveillance satellites and ground radar systems that provide warning of Soviet missile attack. This activity is managed by the "SX" office at SAC headquarters. Another SAC organization, the 1st Strategic Aerospace Division, manages Vandenberg Air Force Base, which will be the launch site for military Shuttle operations. The "1st STRAD" also runs the Defense Department's Defense Meterological Support Program (DMSP) weather satellites and is in charge of the new navigation network, the 18-satellite Global Positioning System (GPS), that is slowly taking shape. Its planning activities for Shuttle operations will be critical to the scheduling and turnaround of military Shuttle flights.

AIR FORCE SYSTEMS COMMAND (AFSC)

This command performs research, development, and acquisition for the Air Force of everything from armaments, radars, electronics, and space systems to new aircraft designs. AFSC has an internal Space Division in Los Angeles, which has a considerable number of program offices handling Air Force space programs. AFSC, however, usually ends up operating as well as developing these programs, even though it is not a fully operational command.

Because it is acquiring an *ad hoc* operational responsibility as well as a research and development responsibility, AFSC has recently established a new office, Deputy Commander for Space Operations (DCSO). The DCSO's primary responsibility is running the Satellite Control Facility—a single facility controlling most Air Force and DOD satellites, as well as several Navy satellites. (Other Navy satellites are controlled from another facility in the United States.)

AEROSPACE DEFENSE COMMAND (ADCOM)

This major command provides many of the assets controlled by the U.S.-Canadian North American Aerospace Defense Command (NORAD). It also has a long range planning staff at ADCOM Headquarters that looks at space defense, antisatellite, and space surveillance operations. In addition, ADCOM/NORAD operates the Cheyenne Mountain Complex, including the Space Defense Operations Center (SPADOC).

* Adapted from a paper submitted by U.S. Rep. Ken Kramer

AIR FORCE COMMUNICATIONS COMMAND (AFCC)

AFCC runs the Air Force communications that are routed through space.

The Air Force headquarters also has several specialized planning offices that develop new missions or technologies for space. The Plans and Operations (Space) Office on the Air Staff was set up September 1, 1981. The group's establishment is the first formal Air Force recognition that the service had to develop an operational approach to space systems that matches what has long existed, as a standard planning perspective, for the aircraft side of the service. In short, it attempts to provide a centralized planning structure for space operations where one did not exist before.

Meanwhile, another separate headquarters staff, Research and Development (Space), conducts its own research activity on space systems. Coming from a research perspective, this organization interacts closely with specialized research organizations, such as Defense Advanced Research Projects Agency (DARPA).

The Deputy Assistant Secretary for Space Plans and Policy is under the Office of the Secretary of the Air Force. This individual is the Air Force's highest ranking civilian official whose primary responsibility is formulating Air Force space policy.

The Air Force also participates in a tri-service planning effort that may be of considerable importance in the future. This program, called TENCAP (Tactical Exploitation of National Capabilities), is designed to extract useful tactical information from surveillance satellites and other sensors that are already operating in space.

Finally, there are three separate offices or organizations within the Air Force that plan or operate classified programs, including those conducted with other agencies or services.

The Navy, meanwhile, has its own recently established Directorate of Space Systems which handles all Navy space activity, including classified programs, communications, and Navy participation in the TENCAP program described above.

The Army's Ballistic Missile Defense Systems Command is developing a considerable "space focus" in its own right. It looks at the deployment of long range antiballistic missiles (ABMs) that reach and intercept Soviet missiles far from the U.S. homeland. It is also developing an optical probe for attack assessment, which will be fired from American territory into any approaching target mass to provide last minute verification and tracking of an attack. The North American Aerospace Defense Command will have operational control of any deployed ABM system (as it did during the Safeguard ABM system's brief life in 1975) and also of the optical probe. Lastly, there is a classified program run by the Army Space Program Office.

DARPA is DOD's primary technical research organization. It is critically involved in advanced space systems research and, as a consequence, is an indirect player in the space policymaking process. In the past several years, DARPA has provided crucial advocacy for directed energy weapons used in antisatellite operations and ballistic missile defense.

The intelligence agencies (CIA, National Security Agency, Defense Intelligence Agency, and other offices) also plan and operate major space surveillance systems.

The Defense Communications Agency (DCA) is in charge of the department wide communications, including those routed through space. DCA coordinates Joint Chiefs of Staff (JCS) operational requirements for communication systems that other organizations must follow in space communications design.

There are several crucial offices that are deeply involved in planning space systems at the higher policy and planning levels that come under the Office of the Secretary of Defense and the assistant secretaries. One such office is the Deputy Undersecretary of Defense for Command, Control, Communications, and Intelligence. This office plans and oversees the electronic systems, including those in space, that control and communicate with our forces or provide the large quantities of intelligence data that come from around the world. There is also a Deputy Undersecretary of Defense for Research and Engineering (Strategic and Theater Nuclear Forces). This office works on strategic attack forces, including laser weapons and theater nuclear forces.

The senior official within the Department of Defense who is charged with formulating and coordinating space policy is the Deputy Director for Intelligence and Space Policy. He reports to an Undersecretary of Defense for Policy, who reports to the Secretary of Defense. Recently, a high level department wide committee was established to coordinate defense space policy and activity: the Defense Space Operations Committee, or DSOC, which is chaired by the Secretary of the Air Force.

All above organizations, commands, agencies, or offices operate within the defense establishment and its associated intelligence agencies.

The civilian space agency, the National Aeronautics and Space Administration, also maintains a close relationship with the military space program, since its Shuttles, upper stages, and other operating elements are often developed and funded cooperatively with DOD.

For example, it was a large military payload and a military performance requirement that determined the size and payload lifting capacity of the Shuttle bay and also determined the Shuttle's ability to fly considerable distances within the atmosphere after reentry (its so-called "cross range" performance).

The executive branch of the government is officially in charge of overall national space policy. This policy is undertaken at the White House level and includes the National Security Council and the President's Science Advisor, who is supported by the Office of Science and Technology Policy. The present Science Advisor, Dr. George Keyworth, is currently conducting a space policy review for the Reagan Administration.

As one can see, the Air Force is by far the service with the most pervasive and diffuse organization for space. It is also the service most in need of reorganization in its space efforts. A recent symposium paper by Dr. Charles W. Cook, the Air Force Deputy Assistant Secretary for Space Plans and Policy, described in grim detail what this organizational fragmentation has meant:

> . . . One of the most serious consequences of the wide distribution of responsibility for space operations (within the Air Force) is the absence of any centralized point within the Air Force for conducting long-range planning for space systems and support functions . . . Planning has been hampered by a lack of vision within the Air Staff and the OJCS (the Office of the Joint Chiefs of Staff) for space operations. It has lagged the potential for using the benefits of space. [NOTE: This paper was written eight months before the establishment of the new Air Force office.]

> . . . Space operations comprise a young but steadily growing mission area. The associated technology, doctrine, and policy are evolving as the potential of space as a medium of warfare becomes increasingly apparent. As our national space policy and doctrine mature, it would be a grave mistake to fragment their growth among several commands . . .

> . . . There is no significant effort underway to develop doctrine, plans and requirements for controlling (space) weapons. Without an assignment of responsibility for operations

planning within a centralized organization, attempts to develop space-based weapons and supporting C^3I (Command, Control, Communications and Intelligence) programs are likely to take longer and be more expensive.

(Dr. Charles W. Cook, "Organization for the Space Force of the Future," drafted January 21, 1981, presented at the Air Force Academy's Military Space Doctrine Symposium, April 1-3, 1981; pp. 479, 481, and 484 of the symposium report, *The Great Frontier: Military Space Doctrine, Volume II*).

Since the delivery of Dr. Cook's pessimistic appraisal, the Air Force has taken action to redress the planning deficiencies he described. Nevertheless, the U.S. military space program needs considerable organizational realignment as we move through the 1980s, and a number of basic organizational and policy decisions remain to be made.

CHAPTER IV: PART C
TECHNOLOGY TRANSFER

INTRODUCTION

It is absolutely imperative that the United States maintain its technological advantage over the Soviet Union. The degree to which the United States can afford to involve other nations, multinational corporations, and even U.S. industry in High Frontier programs is dependent upon assurances that critical, space-associated technologies will not be transferred to the USSR. If the U.S. and its Free World allies are to ensure that the security and economic benefits of space are free from interference of hostile political systems, we must improve our safeguards for our key technologies.

THE SOVIET REACTION

From an historical perspective, it is clear that any shift in the strategic balance in favor of one of the superpowers will provide sufficient incentive for the other superpower to exploit new technology to redress the emerging imbalance. The Soviets, by virtue of their aggressive ideology and resulting geostrategic and geopolitical objectives, have demonstrated greater determination and effort to achieve military superiority over the United States. They are fully aware of the consequences due to changes in the political-strategic environment. It should be evident that they are not going to stand by idly and observe a United States effort to change the strategic equation passively.

A review of Soviet assessments of the Reagan Administration's national security policy clearly reveals a Kremlin understanding that the new government in Washington is systematically undertaking to redress the shift in the "correlation of world forces." It is also clear that Moscow views developments in the military arena as the central issue. The Kremlin categorically insists that the USSR can and will do "everything necessary" to prevent a shift in the present military balance. It confidently asserts that it can succeed in its efforts. Speaking in Kiev on May 9, 1981, Brezhnev said, " . . . if we are compelled, we will find a quick and effective response to any challenge by belligerent imperialism."

Marshal Victor G. Kulikov, the Commander-in-Chief of Warsaw Pact Forces, stated the same thought in the following manner: "The Soviet people . . . will not allow potential enemies to be able to gain the upper hand in any kind of armament or technology."

Considering proven Soviet policies and methods, one can surely expect that the USSR will launch concerted, intensified efforts to obtain from the United States those technologies which will assure the leadership in the Kremlin that the U.S. strategic programs will be neutralized.

It is a well proven fact that Western technology has been used by the Soviets to build up their military capability. A recent report from the Pentagon states, "Without the transfusion of U.S. technology and equipment, the Soviet Union's capabilities would almost certainly have remained at the 10 to 12 year gap of the 1965 era."

The FBI reports that much of the transfer also involved covert operations by the Soviets. Included is "theft of proprietary information at considerable cost, illegal transshipments of our technology to Soviet Bloc countries, penetration of computer systems, and compromise of employees."

The report, authored by FBI Director William Webster, also charges that Soviet officials and visitors came to the United States to attend scientific and professional association symposia only to pilfer information that could be applied for military-strategic and intelligence purposes.

Webster adds, "There is a considerable threat where foreign agents either steal technical information or corrupt an employee to steal this data for money."

The DOD, in its recently issued document, *Soviet Military Power,* states that the industrialized Free World over the past decade has supplied the Soviets with "billions of dollars" worth of "efficient machine tools, transfer lines, chemical plants, precision instrumentation, and associated technologies."

The areas in which the Soviets have made great technical strides with witting or unwitting assistance from the Free World include: directed energy weapons, electronics and computers, explosives, precision weldings, advance composite materials having great strength and low weight, space technology, and others.

TECHNOLOGICAL COMPETITION

With respect to national defense, the term "technological competition" refers to the efforts of competing political-economic systems to maintain, or to achieve, superiority in high technology areas that are important to effective military systems. The history of such a competition between the United States and the USSR dates to 1943 when the Soviet Union began its effort to develop an atomic bomb. The unexpected orbiting of Sputnik by the Soviets in 1957 shocked the United States and for the first time focused broad public attention on the Soviet scientific and technological capabilities and objectives. This event also resulted in a rapid development of our own space science and technology.

The competition between the United States and the Soviet Union continued in all phases of manned and unmanned space programs and in the development of strategic weapon systems. In this era of unprecedented change, our technological strength is the key to our long range survival as a nation.

American security, like the American economy, stands on a foundation of technological superiority. We need superiority in defense technology for two reasons. First, because our open society tells our adversaries what we are planning in military technology while their secrecy forces us to provide for many possibilities. Second, in military operations we traditionally depend on superior quality to compensate for inferior numbers.

The United States continues to hold a technological lead over the Soviet Union in many critical areas that are vital to our national security, but that lead has been diminishing. In some very important areas it is gone; in others the Soviets are ahead.

Moreover, the technology balance is dynamic. In examining the current technology balance and its dynamics, qualified analysts agree that the USSR expends a very large and determined effort and that they are inexorably increasing their level of technology relative to ours. In fact, they are seizing the initiative in many important areas, such as high energy laser beam and charged-particle beam weapons, surface-effect vehicles, and antipersonnel pressure weapons.

Technological development is molding future Soviet strategy. From all indications, the Soviet strategy will continue to center on world dominance, with technology as a key factor. A crucial element in our strategy of deterrence is the maintenance of a margin of military advantage through possession of a number of sophisticated technologies.

PROTECTION OF HIGH FRONTIER TECHNOLOGY

Any thrust toward deployment of advanced technological systems requires parallel efforts to protect against the overt and covert leakage of those technologies to potential enemies.

The prevention of military and military related technologies (''dual purpose'' technologies) exports from the United States to potential adversaries is a national security imperative. We must use all legal methods to protect our technological lead and not contribute to the military strength of potential adversaries.

Since the end of World War II, the United States has relied principally on two laws to control export of certain goods and services to all destinations including to our principal adversaries. They were the Export Control Act of 1949 and the Mutual Defense Assistance Control Act of 1951, also known as the ''Battle Act.'' These statutes, enacted during the height of the U.S. policy of containment, reflected the prevailing view at the time that any Soviet or Eastern European economic development would ultimately contribute to the military capability of the Soviet Union and its allies, and was therefore to be inhibited, discouraged, or forbidden outright. Accordingly, the sale of nearly all products or technology to the Soviet Union, its European allies, and the People's Republic of China (PRC) was embargoed.

During the 1960s and 1970s, U.S. foreign policy, including foreign trade policy, underwent a significant change. The pressures of business and other interests substantially increased for the liberalization of trade with communist countries. This resulted in changing the substance of the Export Control Act. In its place, in 1969, the Congress enacted the Export Administration Act. This statute, premised on the belief that increased trade with the Soviet Union and its allies might ameliorate broader political conflicts, facilitated substantial increase in East-West trade. The result was the ''hemorrhaging'' of U.S. technology, which contributed heavily to the dramatic growth of the Soviet military power. It also brought about public realization of the problem as well as a shift in attitudes among U.S. legislators. In 1979, the Congress enacted the Export Administration Act of 1979, which was a step in the right direction, but is insufficient to redress the existing problem. The significant and positive new provision of the Act was a mandate to DOD to develop an initial list of militarily critical technologies (MCTL) to be incorporated into the Commodity Control List (CCL), but at the discretion of administrative authority at the Department of Commerce. The MCTL has yet to be incorporated in part or as a whole into the Commerce Department's CCL.

An available alternative is to be found in the Arms Export Control Act of 1976 (AECA). This governs sales of technological data and hardware placed on the U.S. Munitions List. The Act relates to foreign military sales and licensed production or coproduction and recoupment of the U.S. military research and development costs. The intent of the Act is to retard the transfer of weapons manufacturing capability from the United States to other nations. The AECA may be suitable for the purpose of safeguarding High Frontier core technology in an international environment.

The participation of U.S. allies in the High Frontier's nonmilitary effort would add an international dimension. The nations of NATO have a tremendous economic and strategic incentive to cooperate. It is clear that our allies, in order to share the benefits, must accept the safeguards necessary to protect High Frontier core technology and its space applications.

In that respect, a mechanism to achieve protection of core technology in an international environment is the Arms Export Control Act and the U.S. Munitions List. The AECA is most suitable and efficient because it provides adequate protection and, at the same time, the mechanism is established, familiar, accepted by the NATO alliance, and provides for transfer, handling, and safeguards of military technology within the alliance. In other words, neither the procedures nor the relevant legal and security obligations represent something new which would require adjustment.

Although the transfer of technology to communist countries has become a contentious issue recently, both within the United States and among NATO partners, technologies on the U.S. Munitions List were not the subject of any dispute. Both classified and unclassified technical data and hardware of U.S. origin have been transferred to the NATO allies without significant difficulties. There are no complaints as to the terms and conditions which accompanied such transfers.

THE ARMS EXPORT CONTROL ACT AND MUNITIONS LIST

The provisions of AECA distinguish between classified and unclassified technical data and hardware. Nevertheless, any item placed on the Munitions List (ML) requires an export license which specifies policy conditions and technical level conditions of transfer. The constraints on the recipient are part of the license agreement; e.g., a German firm party to the agreement assumes the condition that the technology in question can be retransferred only to specific countries within the NATO alliance and *not* to those outside of the alliance. Under provisions of the AECA, all items are, without exception, reviewed by DOD. Furthermore, every item on the ML is automatically embargoed for export or reexport to a communist country. Lastly, one can track the movements of data and hardware.

Also, other provisions of the Act which have not been questioned by our allies are that the United States has a free hand to write the terms of any transfer of technical data or hardware. Any transfer of classified items requires the recipient party to establish security procedures which are required of U.S. defense contractors handling classified data and hardware.

In the United States, if a particular technology is placed on the ML, the owner of such technology is required to register with the Department of State, and to establish security procedures required by DOD's Industrial Security Manual. Under the DOD security provisions for those who handle classified technical data and hardware, the subject must have his facilities certified as secure and the participating personnel cleared for the handling of classified items. The Industrial Security Manual also stipulates that each firm handling classified material must have its own security officers who must work together with the Defense Industrial Security Clearance Office, which provides security clearance and maintains files on individuals in industry with such clearances.

Much of High Frontier core technology would be classified. All security criteria which apply to domestic firms would also apply to foreign participants. There DOD, DIA, and CIA should prepare status reports on the adequacy and effectiveness of the security in a participating country or its industry.

It should be reemphasized that U.S. NATO allies have accepted the aforementioned criteria, conditions, and safeguard procedures. Consequently, we do not expect any significant problem in that respect with regard to protection of High Frontier core technology from Soviet penetration.

THE FRAME OF REFERENCE

For our purposes, the technology considered for protection must include arrays of technical information and know-how, keystone equipment and materials, and the goods accompanied by these. The criticality assessment of a relevant technology must be based on its strategic utility. Upon determination of which technology must be controlled, such technology must be placed on the Munitions List, which is the only way to guarantee the fulfillment of requirements under national security considerations. By placing the High Frontier core technology on the Munitions List, we are also protecting leaks of the same, via third party, because the U.S. will be the sole authority.

Special attention must be given to a particular core technology which is the key to the High Frontier concept. U.S. and allied microelectronic is imperative for the successful implementation of High Frontier. It is present in every aspect of core technology, and it is relevant to every system of the High Frontier concept, military or nonmilitary.

Microelectronic is one of the main technological advantages the United States enjoys over the Soviet Union. It permits the reduction in size and weight of every relevant piece of space hardware and, at the same time, provides for optimization of capabilities and performance in relation to size and weight. It is estimated that the United States has a lead of some seven to ten years over the USSR in microtechnology.

CHAPTER V: URGENT REQUIREMENTS AND COSTS

The preceding chapters present the case for High Frontier military and nonmilitary efforts in space and those nonspace efforts essential to the change of strategy involved. This chapter addresses immediate and urgent requirements inherent in the concept and provides an illustrative set of programs to meet them.

MILITARY REQUIREMENTS

- A point defense for U.S. ICBM silos which can within two or three years at a cost less than that of super hardening, destroy any confidence the Soviets might have in a first strike against our deterrent.
- A first generation spaceborne ballistic missile defense, deployable in five to six years, capable of significant attrition of a Soviet missile attack in the early part of the trajectory.
- A second generation space defense system, deployable within 10 or 12 years, capable of attacking hostile objects anywhere in space with advanced weaponry.
- A manned military space vehicle, deployable within three to four years, capable of surveillance, inspection, on-orbit maintenance, and space transportation missions to support BMD and to enhance near term space control.
- A civil defense program of sufficient scope and funding to take advantage of the proposed active strategic defenses, adding to U.S. deterrent strength. An effective, low cost civil defense program is described in Appendix F.

COMBINED MILITARY AND NONMILITARY REQUIREMENTS

- An improved space transportation system designed to lower the cost-per-pound in orbit to under $100.
- A low Earth orbit space station to support military and nonmilitary missions.

NONMILITARY REQUIREMENTS

- A vigorous developmental program to encourage space industrialization, including solar power satellite technology.

AN ILLUSTRATIVE SET OF HIGH FRONTIER PROGRAMS

There are several options available to satisfy the urgent requirements set forth above. To prove the basic feasibility of the High Frontier concept and to establish time and money parameters required to implement the concept, an illustrative set of systems that can meet the urgent requirements was analyzed in greater detail and is described below.

Each program is described in greater detail elsewhere in the study. Cost estimates are in constant dollars. The estimated time required to reach the initial operational capability as well as cost estimates are critically dependent upon a special management system to minimize bureaucratic delay. This management system is discussed in Chapter VII and Appendix H.

QUICKLY DEPLOYABLE POINT DEFENSE

The key to any effective point defense system is solution of the radar vulnerability problem as discussed in Chapter IV, Annex A. Once that is accomplished, any of several kill mechanisms can meet the restricted point defense requirement of High Frontier.

A partially tested system exists that could meet the requirement to destroy Soviet confidence in a first strike against our own missile silos. It is a simple system that fires a large number of small conventional projectiles, which form a barrier against a warhead approaching a U.S. missile silo at about one mile from the target. It could be described as "dynamic hardening" instead of as an antimissile system. If deployed to intercept only the first Soviet warhead approaching a silo, it would cost $2-3 million per defended silo. If it is to intercept a second warhead, the costs increase to about $5 million per silo. (See Appendix B for a full discussion of this system).

FIRST GENERATION SPACEBORNE DEFENSE

The requirement for an initial spaceborne ballistic missile defense system can be met by using off-the-shelf hardware to create a multiple vehicle, orbiting system. This system would deploy non-nuclear kill vehicles to destroy Soviet missiles in the early phase of trajectory (Figure 11). Enough weapons-carrying satellites would be orbited to ensure continuous coverage of Soviet ballistic missile trajectories, including those of SS-20 Euro-strategic missiles and submarine launched missiles. This system could provide protection to the allies as well as to the United States. (See Appendix C for a fuller discussion of this system).

The multiple satellite deployment permits one satellite to defend itself and several others from hostile attack. It also has the potential for forming the basis of a highly effective and secure command, control, and communications (C^3) system. Since the system makes maximum use of off-the-shelf space hardware components, it may be the least expensive and quickest available option. Deployment of this system could begin in as little as three years and be fully deployed in five or six years at a minimum cost of some $10-15 billion.

The size and weight of the carrier satellite in this system is constrained by the limitations in throw weight of the MX booster which is proposed as the method of insertion into orbit. Variations on this concept using larger satellites with greater capabilities placed in orbit by Shuttle or larger boosters may prove technically preferable. These variations would raise costs to about $40 billion and extend full operational capability dates by about two years.

SECOND GENERATION SPACEBORNE DEFENSE

The most promising possibility for a second generation spaceborne defense is product improvement of GBMD I. With the addition of advanced infrared sensing devices the first generation can be made capable of attacking individual warheads throughout their trajectory up to reentry into the atmosphere. This system could be ready for deployment in 1990 at a cost of about a $5 billion add-on to GBMD I costs.

The requirement for higher technology space defense systems might also be met by a high powered laser system on the ground with redirecting mirrors on satellites (Figure 12) or by beam weapon systems deployed in space or in pop-up installations on the ground. These systems are currently being researched. Costs to continue research should probably be increased by about $100 million per year. (See Appendix E for a fuller discussion of second generation systems.)

HIGH PERFORMANCE SPACEPLANE

There is an urgent need to develop a multipurpose, manned space vehicle to perform a wide variety of missions such as inspection of friendly

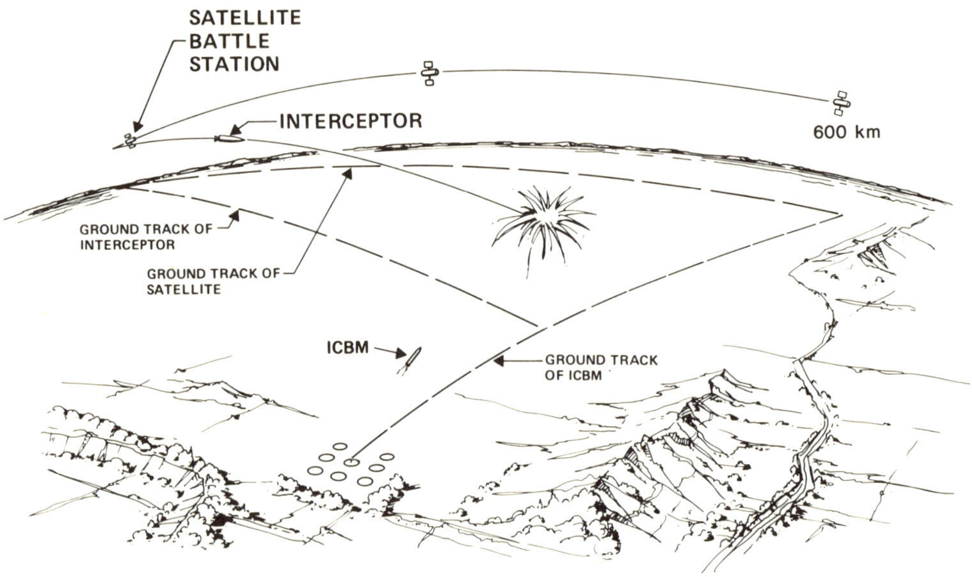

Figure 11. Boost Phase Intercept, Geometry

This system uses multiple satellites armed with small conventionally-armed sub-missiles to destroy ICBMs and other strategic missiles shortly after their launch. It uses off-the-shelf technology and can be deployed in less than five years.

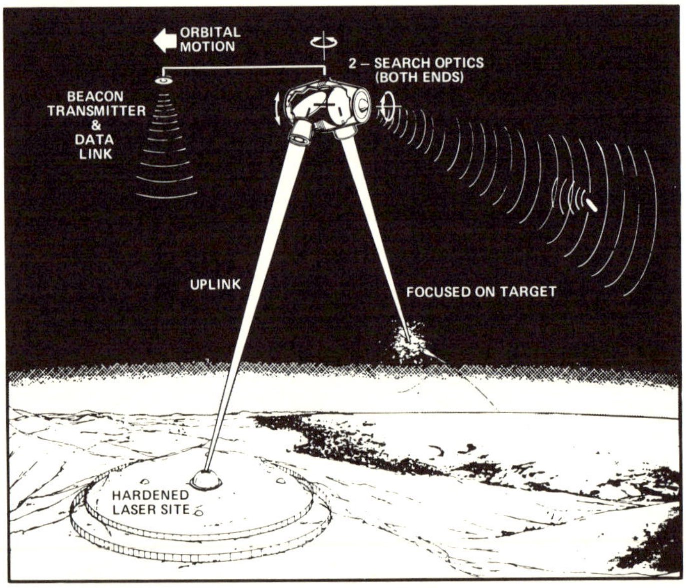

Figure 12. Advanced Global Ballistic Missile Defense

One example of a second generation space-defense system using a laser weapon.

or suspect space objects, satellite and space station protection, and adjustment or retrieval of satellites. One such vehicle is the high performance spaceplane, or one man space cruiser which uses available hardware components and technology and which could be operational in several years for less than $500 million. (See Appendix D for fuller discussion of this system.)

CIVIL DEFENSE

Civil defense is a multifaceted endeavor, in which the utility and cost effectiveness increase sharply, when considered in conjunction with active defenses. (See Appendix F for a fuller discussion.)

IMPROVED SPACE TRANSPORTATION

The immediate answer to improved space transportation is an upgrade of the current Shuttle program to improve turnaround time and to create an unmanned cargo only version. At the same time, development work should begin on a vehicle with a much heavier lift capability. These programs would cost an estimated $6 billion over a 10-year period.

MANNED LOW EARTH ORBIT SPACE STATION

A currently proposed military Space Operation Center should be given high priority and expanded in concept to include provision for fly along industrial/commercial space installations. The space station should be equipped to receive energy for operations from a prototype solar power satellite. A 10-year program to deploy this space station should cost about $12 billion.

A SPACE POWER SYSTEM

This requirement can be met by a proposal using known technology that would place a solar power satellite in geosynchronous orbit and place a microwave receiving antenna and conversion system on Earth, providing 500 megawatts of continuous electrical power. This pilot system, modified to include a capability to provide power to a space station with laser transmission, would cost about $13 billion.

SPACE INDUSTRIAL SYSTEMS RESEARCH AND DEVELOPMENT

The costs of R&D for industrial space applications would probably be borne almost entirely by interested private enterprise, with little more than $50 million per year in government support.

COST CONSIDERATIONS

The options available to meet High Frontier objectives involve both space and nonspace systems, conventional and nuclear, and various types of beam weapons. Selection should be a function of the Systems Selection Task Force recommended in Chapter VII. Actual costs cannot be determined until specific systems are selected; but, for the purposes of this study, we have roughly estimated the costs of military and nonmilitary illustrative programs.

A combination of military and civilian programs which cost about $20 billion over five years and about $35 billion through 1990 should meet the concept's requirements. A careful consideration of tradeoffs, in terms of current DOD requirements that High Frontier military systems would meet partially or fully, indicates that there will be little or no increase in projected defense budgets. Commercial and allied support of High Frontier nonmilitary systems could further reduce demands on the U.S. Federal budget.

Cost data validity for the illustrative High Frontier programs range from hard to very soft. In all cases, proponents of the programs (private companies or concept authors) provided the data. This raises a possibility of bias on the low side of cost projections. Our analysis took this into account by rather arbitarily increasing (in some cases as much as 100 percent) the figures provided by basic sources. This was done despite the fact that we foresee major savings in estimated costs if the acquisition measures proposed in Chapter VII are implemented.

The costs estimated for several programs are much firmer for program totals than for year-to-year breakdowns. Further, the cost calculations take no account of inflation and appear in constant dollars. (See Figure 13, High Frontier Cost Parameters.)

Despite uncertainties as to details within this projection, it nevertheless demonstrates the feasibility of pursuing a vigorous set of programs at reasonable costs. The estimates in Figure 13 indicate that the first five years would cost $24 billion; about $18 billion for military and $6 billion for nonmilitary systems. Over eight years, the total costs reach about $40 billion; $27 billion for military, $13 billion for nonmilitary. Beyond eight years, roughly $10 billion more may be required to operate and maintain deployed systems and to reach initial operating capability for an advanced Shuttle, a pilot solar power satellite, and an advanced space defense system. Thus, total commitment to this set of High Frontier programs could amount to some $50 billion over the next 10 to 12 years.

These figures may well err seriously on the low side, but the military and economic consequences of success would make these programs a strategic bargain at *twice* the indicated costs. Alternate GBMD systems studied called for additional costs, which, however, do not appear to double the estimates put forward in this study. The amounts used above are limited to acquisition and operation of systems. In overall strategic military cost it is important to include the value of defensive systems as an "insurance policy" for defense expenditures already made and programmed for the next decade. This analysis in Appendix A offers a different perspective which makes the High Frontier programs even more attractive.

COST OFFSETS

We can also estimate orders of magnitude offsetting savings to come from capabilities achieved by High Frontier programs. Clearly, costs are in the ballpark of the feasible. They offer more potential than some programs now getting surprisingly high long range "seed money."

The basic cost question should always be one of "opportunity cost," that is, cost of the best alternative opportunity foregone. Ideally, economists select the lowest cost alternative for reaching any given objective. In practice it is seldom, if ever, feasible to describe alternatives of literal equivalence (in the military field, generally called effectiveness; in civilian endeavors, benefits).

High Frontier military programs could be initiated with funds now earmarked for the survivability of strategic systems. Another source could be the reprogramming of funds earmarked for systems or support activities that would not be required once the High Frontier's strategic defense capabilities were in place.

Not yet knowing the High Frontier specific systems or their initial operational dates, one cannot suggest reprogramming on a year-by-year basis. However, the Administration's decision on the B-1 and MX included a commitment to research and development of defensive systems to protect our ICBM capabilities. Presumably it included necessary funds for this, and the implementation of High Frontier systems would satisfy this Administration goal.

With respect to reprogramming, we assert that the proposed global ballistic missile defense capabilities would reduce, or eliminate entirely, the need to harden existing silos to accommodate MX missiles. Funds designated to superharden the silos could be diverted to spaceborne defense

High Frontier Systems	IOC	FY1	FY2	FY3	FY4	FY5	5-yr total	(Military)	(Civilian)	FY6	FY7	FY8
Point defense of 200 silos	2 yr	150	1,040	30	30	30	1,280	(1,280)	(0)	30	30	30
Global BMD I (off the shelf)	5 yr	300	1,000	3,000	3,000	3,000	10,300	(10,300)	(0)	3,000	60	60
Global BMD II (advanced technology)	10 yr	50	50	100	300	1,000	1,500	(1,500)	(0)	1,000	1,000	1,000
High performance spaceplane (includes ACFT Launch)	4 yr	10	50	100	150	200	510	(510)	(0)	200	200	200
Shuttle improvement	10 yr	200	300	500	500	500	2,000	(1,000)	(1,000)	500	1,000	1,000
Space tug-orbital transfer vehicle	8 yr	16	56	234	325	325	1,156	(0)	(1,156)	253	205	200
Low-Earth orbit space station (military and civilian)	8 yr	37	138	1,364	2,946	2,337	6,822	(3,411)	(3,411)	2,023	800	200
Satellite power system (R&D plus pilot)	12 yr	30	60	100	150	200	540	(0)	(540)	200	1,300	1,600
Total by year (rounded)		800	2,700	5,400	7,400	7,600	24,000	(18,000)	(6,000)	7,200	4,600	4,300

Note: Millions of constant dollars

Figure 13. High Frontier Estimated Cost Parameters

systems and to the point defense of ICBM silos recommended in this report.

Pursuit of Assured Survival by a combination of space based defense systems and point defense should solve the strategic force survivability problem. Without strategic defense this country is left with only two options in the face of Soviet nuclear force. We can resign ourselves to the calamitous absorption of a surprise first strike or depend upon our technical surveillance systems to let us launch on warning or as the warheads arrive on target. The gloomy consequences of these two options and the dramatic results when various defensive systems are added are described in Appendix A.

Funds now programmed to seek other solutions to this problem, such as MX deployments, rebasing the B-52 force, or pursuing alternative ICBM passive protection schemes, could logically be reallocated to the active defense programs proposed by High Frontier.

For nonmilitary or dual purpose systems, such as a manned space station and a pilot solar power satellite (SPS), development and acquisition funding should be provided largely from nondefense sources. Solar power satellites offer an attractive prospect for large scale free enterprise investment in a potentially profitable space activity. While the costs associated with building and deploying an SPS network are sufficiently sizeable to require direct government participation, industry interest and investments in SPS research might be encouraged if government guarantees were made available.

Offsets in Competitive Systems

Other programs not clearly seen as alternatives to a given proposal but subject to possible reduction or elimination deserve attention. (For example, an effective BMD might permit deployment of fewer and less hard ICBMs.) Clearly, the costs of such indirect alternatives should be deducted or included in negative terms in the opportunity cost equation. Thus, the opportunity cost of some of the proposed High Frontier systems may turn out to be small or even negative.

Point Defense

We believe that a defensive system for hard sites (ICBM silos) could be deployed in about two or three years at most. It would use nonnuclear kill mechanisms, permitting distributed defense systems with intercepts at a few kilometers slant range. One such system, the SWARMJET, has been estimated to cost on the order of $2.5 to $5 million per defended point, or about $1 billion for 200 MX or Minuteman III silos. Other low cost, quick systems have been proposed. The costs involved in closing the land based ICBM window of vulnerability, at least partially, might well come from deferring MX silo hardening pending future decisions on ultimate MX deployment modes.

Global Ballistic Missile Defense (GBMD)

The system described in Appendix C could be available in five or six years. Its off-the-shelf component cost of $12.6 billion is remarkably low. The approximately $15 billion, eight year estimate provides for unforeseen costs of ground control, replacements, etc. The system's first payoff would be to filter Soviet ICBM attacks in early trajectory. The combined filter effect of silo point defense and a spaceborne defense in depth can be very high. Even if each system is "leaky," i.e., if the average single shot kill probability (SSKP) is lower than hoped for, the two systems together would probably be highly effective. For example, even if each system has a 30 percent leakage rate, together they will still save 91 percent of the protected force. If the leakage is as high as 50 percent for each, together they will save 75 percent of the force.

A boost phase intercept system (like a preemptive counterforce attack) also degrades all other enemy missile attacks. For example, to the extent that a Soviet attack might have included C^3 targets, it may be possible for the United States to save a significant part of the $18 billion recently proposed for C^3I improvements. In fact, the multiple satellites of the defense system may be able to enhance C^3 survivability by hosting a redundant and hard to interdict space communications relay system. It might also make possible a reduction in the programs under development for the wartime continuity of government, including the protection of presidential successors and their support teams. To the extent that the Soviets might attack military and economic targets, there could be offsetting savings in the hardening and dispersal requirements. There is also the possibility that proposed (not yet implemented) U.S. civil defense programs of $1 billion per year or more might be reduced substantially in the future if the active defense systems of High Frontier are deployed.

Our allies could also be protected by the proposed spaceborne defense systems as a result of its potential effectiveness against SS-20 intermediate range attacks. This could simplify the impending theater nuclear force negotiations. The GBMD would strengthen—or rather restore—the U.S. strategic umbrella. Such a system might contribute more to the reduction of nuclear proliferation incentives than all the fuel cycle controls, international inspection teams, and nonproliferation treaty measures taken together.

None of the High Frontier proposals protects against a depressed trajectory attack by Soviet submarine launched missiles against strategic air bases. However, the vulnerability of these bases to such attack becomes less critical if our ICBM force is no longer vulnerable to a first strike. This fact could affect the costly proposal to move bomber bases inland as well as other reduction requirements designed to reduce bomber force vulnerability.

High Performance Spaceplane (Space Cruiser)

The space cruiser could be operational in three to four years. Its R&D costs should be relatively modest because most of the components would be off-the-shelf and principal development costs would be in systems engineering. The savings provided by the space cruiser might include reduction in the number of standby vehicles since the space cruiser could provide repair in space. There may also be savings in the size of various space vehicles because of the refueling potential of a space cruiser. Of special interest might be the facilitation of crew change as well as maintenance and supply for a low orbit manned space station. Within the illustrative High Frontier programs, there may be a cost tradeoff between the function of the space cruiser and the orbital transfer vehicle (OTV).

CHAPTER VI: IMPACTS

It is difficult to overstate the potential impact of a U.S. commitment to the High Frontier concepts. Together, the programs would reverse several key trends which run counter to the interests of the U.S. and the Free World. These include the military balance, the energy problem, the dismal prospects for underdeveloped nations, and the general erosion of spirit in the West. Much of the beneficial impact of High Frontier can reasonably be expected to occur long before the first defensive satellite is on station or the first kilowatt of power is delivered from space. The mere announcement of a cohesive U.S. strategy designed to discard menacing balance of terror military doctrines and open new horizons for human enterprise might trigger a wave of general optimism in the Free World. In any case, it would restore the image of the United States as the purposeful leader of the West.

MILITARY IMPACTS

High Frontier military programs would have profound strategic impact. They move the U.S. away from the unstable world environment of balance of terror (Mutual Assured Destruction) toward one of Assured Survival. The change of emphasis away from total reliance on nuclear retaliation would provide responses to Soviet threats to U.S. and allied security other than the amassing of even larger stockpiles of ever more expensive offensive nuclear weapons.

The principal military impact of High Frontier will be the achievement of President Reagan's desired "margin of safety" in the quickest, most economical way. The U.S. cannot afford to engage in a quantitative arms race with the Soviets in offensive nuclear weapons only, nor can the U.S. afford the costs of an effective system of deceptive deployment for land based missiles (e.g., MX-Racetrack).

A new balance between strategic defensive and offensive systems would by no means obviate the need for retaliatory systems. Indeed, no one can reasonably postulate a strategic defensive system so effective that it would prevent severe damage from a nuclear retaliatory attack. Certainly the defensive systems proposed by High Frontier would not prevent *retaliation* by the Soviets using submarines, bombers, or cruise missiles, although they would put a Soviet *first strike* out of the realm of rationality.

The same is true of Soviet strategic defenses. While they can sharply reduce the effects of a U.S. retaliatory attack, they cannot prevent it. Soviet strategic and civil defenses constitute a menace today only because they exist in a strategic framework in which the U.S. as a whole and its nuclear deterrent specifically are defenseless. This situation provides the rationale, if not the temptation, for a Soviet first strike or nuclear blackmail. On the other hand, the Soviets might not remain deterred if the United States created only a strong strategic defense and allowed its deterrent offensive force to weaken.

Thus the need for credible retaliatory offensive forces remains. However, by creating a balance between protective and striking power, we broaden our options in strategic offensive systems.

Much of the imbalance in strategic offensive systems derives from the fact that the bulk of the Soviet strategic nuclear force is in land based ballistic missiles which can reach their targets in the United States within 35 minutes after a decision to attack. By contrast, two-thirds of the U.S. force capability is in bombers and submarines which require several hours—even days—to respond. Thus the U.S. force is almost totally retaliatory (second strike) in nature. The Soviets have a dangerous advantage in preemptive first strike capability provided by their ICBM forces. If this preemptive threat were neutralized by strategic defenses, the slower reacting U.S. retaliatory weapons would regain their full credibility as components of the strategic forces.

In the absence of strategic defenses, the only ready response to increased levels of the Soviet first strike threat is to add more offensive weaponry and invest in costly hardening and dispersal measures. With a U.S. strategic defense capable of reducing the weight of a Soviet missile attack, thus shouldering much of the damage limiting load, the imperative for counterforce offensive weapons becomes less pressing.

In such circumstances, slower acting, more purely retaliatory weapons such as cruise missiles become more attractive. Further, deployment of additional nuclear offensive weaponry overseas as a response to such threats as the Soviet intermediate and medium range missiles opposite NATO becomes only one of the options available, not the sole option short of acquiescence.

Perhaps the most important impact of the High Frontier concept on the U.S. military establishment as a whole is the restoration of the traditional American military ethic. Over the years, the American soldier's role as defender of the country has been the tie that bound him to a generally supportive citizenry. Strategies of the past 15 years, such as Mutual Assured Destruction, have denied the validity of that role and weakened the bond between the American military man and his fellow citizens. A commitment to a new strategy which restores the fundamental mission of protection of the Nation would once again be consistent with the military rationale of the average American citizen. This would do much to rekindle the willingness to serve in the military and to ease the serious problems facing our Armed Services as a whole.

POLITICAL IMPACTS

The potential for public support of this concept is enormous. If the military and nonmilitary aspects of High Frontier are effectively harnessed together, broad segments of the U.S. body politic are likely to rally in support. Recent elections have demonstrated the widespread desire for improved national security. There is a remarkably large and enthusiastic support base for space based strategic solutions, primarily among younger people. And there is general public disillusionment with the doctrines and strategies of the past.

Adoption of the High Frontier concept could even convert or confuse some of the traditional opponents of defense efforts and technological innovations. It is harder to oppose nonnuclear defensive systems than nuclear offensive systems. It is almost impossible to argue effectively for a perpetual balance of terror strategy if it can be negated by new policies. It is hard to make environmentalist cases against space systems.

Even those naysayers whose basic concern is disarmament will be hard pressed to make a case against High Frontier, the ABM Treaty notwithstanding. It is not necessary to abrogate the ABM Treaty to commit to High Frontier programs.

The proposed spaceborne defensive systems fall into the category described in the treaty as "systems based on other principles" which are "subject to discussion" with the Soviets. Point defense systems can be selected which are so different from ABM systems, as defined in the treaty, that they too could be considered as outside the treaty. Some silo-defense systems could be considered "dynamic hardening"—a substitute for reinforced concrete—rather than an ABM system. Further, the current ABM Treaty is scheduled for review in 1982, and the United States can propose any amendments deemed necessary to accommodate strategic defensive decisions.

A U.S. commitment to the High Frontier concept does not necessitate rejection of arms negotiations with the Soviets. It does, however, mean that future negotiations would proceed on a different philosophical basis. Rather than continuing to pursue agreements which attempt to perpetuate a balance of terror, our negotiating efforts would be dedicated to achieving a stable world of Mutual Assured Survival.

ECONOMIC IMPACTS

There can be little doubt that a strong U.S. commitment to the High Frontier programs would have highly beneficial economic impacts. Some of these will affect the U.S. economy in the near term, primarily through the stimulus to investment in high technology sectors of industry and a probable upswing in general confidence. An increase of 200,000 jobs in the near term as a result of a strong commitment to space has been estimated. Longer term impacts will depend on the rate at which industrial applications are realized and on unpredictable technological spinoffs from the space effort.

One area of commercial space application is already paying its way very well. Space communications is a $500 million-per-year enterprise and growing rapidly. By 1990 it should become a multibillion dollar-per-year industry.

As other industrial applications in space are realized, the total revenues from space industries might reach levels of several tens of billions of dollars per year by the year 2000.

Some of the most beneficial economic impacts of a strong High Frontier effort are indirect and unquantifiable. The demand for highly skilled workers is certain to have an impact on the education system and on the labor market. New products, tools, and services will be required by an expanding space effort. Research efforts will intensify.

Overall, the economic benefits of a strong U.S. commitment to the exploitation of space for both security and industry are potentially very great, but they are no more predictable today than were the future economic benefits of aviation in the 1920s.

FOREIGN IMPACTS

The positive political effects in the U.S. will probably be reflected overseas among our allies. The announcement of a commitment to the High Frontier concepts could have a strong countereffect on the current highly disruptive, ''anti-nuclear,'' or ''peace'' movements in Europe. A bold U.S. strategic initiative would certainly bolster the morale of pro-U.S. elements. The High Frontier concept can add new strength to Free World alliances, making them global rather than regional.

A shared U.S.-Allied commitment to the harnessing of solar power from space could have highly beneficial impacts on foreign relations. If the prospects were good for future supplies of energy independent of the geographical location of fossil fuels, the overdependence of the industrialized West on oil and gas producing countries could be rectified. Further, the prospects for overcoming the presently intractable problems of the underdeveloped nations with space technologies such as solar power satellites could have a beneficial impact on the attitudes of the Third World.

THE SOVIET REACTION

As for the Soviets, their reaction is easily predictable as hostile. They have already moved to counter the U.S. potential to adopt available military space options. They have introduced in the U N (and garnered some support for it among our allies) a new treaty which would ban *all* (not just nuclear) weapons in space. Meanwhile, evidence mounts that they are already in violation of their own cynical proposition. We can expect an extraordinarily strong Soviet propaganda effort against a U.S. commitment to the

High Frontier concepts, including threats of counteraction. However, in both particulars Moscow will find, for substantive reasons, an attack on the High Frontier concepts much more difficult to conduct than past anti-U.S. campaigns.

A High Frontier decision by the U.S., if backed by effective implementation efforts, would severely impact the Soviet Union, perhaps decisively. The Soviet leadership would consider it as offering the best chance, if not the only chance, for realization of President Reagan's stated intention to refurbish U.S. military power to the point necessary for an effective foreign policy.

These conclusions are not assumptions or conjectures. They are the product of exhaustive examination and analysis of what the Soviets themselves say about the situation and issues that would be raised by a U.S. armaments effort on the order of High Frontier.

On the basis of testimony emanating from Moscow, in Kremlin eyes a credible U.S. commitment to High Frontier would:

- Confront the USSR with precisely the sort of armaments competition that the Soviet leadership most fears and is most anxious to avoid.
- Severely tax, perhaps to the point of disruption, the already strained Soviet technological and industrial resources.
- Seriously threaten the very foundations of the strategic structure the USSR has built at great cost over the past 20 years.
- Undercut the foundation for Kremlin claims that the ''correlation of world forces'' has irreversibly shifted in favor of the USSR.
- Force a return to the drawing board to restructure the doctrinal concepts and strategic designs that have been developed in recent years for Soviet victory over the West, whether by nuclear war or by means short of nuclear war.

The Kremlin has always viewed space in terms of its military utility and has recognized that the best opportunity for a decisive U.S. or Soviet breakthrough is in space. Even as the USSR exploited its unexpected dividends from the worldwide psychological impact of the first Sputnik, Soviet attention and interest were concentrated not on the Sputnik phenomenon but on giving substance to the nuclear rocketry strategy that was then emerging from the ''revolution in military affairs'' that had been effected in the USSR.

Khrushchev spoke with suprising candor to President Kennedy about the actual situation in the USSR at their meeting in Vienna in June 1961. He told Kennedy that space cooperation was ''impossible until there is disarmament'' because of the intimate relationship between space and military activities. He said that there had been few ''practical uses of outer space launchings'' which were ''primarily for prestige purposes,'' and that such endeavors as an attempted ''flight to the Moon'' might weaken Soviet ''defense'' efforts.

While the USSR continued to expand its space program steadily in the last years of Khrushchev's rule and thereafter under Brezhnev, the emphasis was overwhelmingly on near Earth activities. These related directly to military objectives and primarily to those that looked toward achieving a system of manned space stations or, as Brezhnev called them, ''cosmodromes.''

Meanwhile, the Soviets have, as a standard practice, charged that U.S. space activities are directed toward military ends. This has been good propaganda from Moscow's standpoint, serving not only to further its efforts to stamp the U.S. with an indelible threat to peace image, but also to blunt the impact of U.S. successes in space. At the same time this charge provided a backdrop for incessant claims that the USSR pursues solely peaceful purposes in space.

However, more is reflected in such charges than propaganda. A genuine fear has been implicitly reflected in Soviet statements that the U.S. might use its technological prowess to best the USSR in using space for military purposes.

In countless ways the Soviets make clear that they are keenly aware of U.S. capabilities to accomplish even the most difficult tasks when it seriously sets itself to those tasks. They do not attempt to disguise their respect for U.S. technological competence. Indeed it can be safely said that in the area of technology, including especially the high technology required for advanced weapons, the Soviets perceive their greatest single vulnerability vis-a-vis the United States.

At the same time, Soviet authorities up to and including Brezhnev have repeatedly warned of the possibilities and far reaching consequences of a weapons breakthrough. In recent years Moscow has strongly pressed a propaganda-diplomatic campaign, first launched by Brezhnev in June 1975, to secure international agreement to place "a ban on manufacturing new categories of mass destruction weapons and new systems of weapons."

The truly massive campaign the Kremlin is conducting against President Reagan's armaments plans is heavily focused on the new weapons issue, with space possibilities as the prime object of attention. The first successful flight of the U.S. Space Shuttle in April 1981 has served as the point of departure for literally hundreds of authoritative Soviet commentaries on alleged U.S. preparations to deploy in near Earth orbits new generations of space weaponry designed, as *Izvestiia* recently reported, "for carrying out strikes against targets in space, the atmosphere, and on the Earth . . . in the hope that the U.S. will be able to avoid nuclear retaliation."

Thus the signposts have already been set regarding the Kremlin's reaction to adoption of a High Frontier strategy by the United States. It would view the move as directed toward what the Soviets themselves have characterized as a possible "absolute weapon" capable of ensuring "invincibility" of the U.S. from missile attacks. While the Kremlin would naturally consider fulfillment of this aim as some years away, at best, knowing the state of the technological art involved, it would still allow that substantial U.S. capabilities could be in place within a relatively short term of two to four years.

The question consequently arises as to what the USSR would do about the new situation created by a U.S. commitment to High Frontier concepts.

One step is certain: Moscow would pull out all stops in its political propaganda maneuvering to deter or deflect the U.S. decision. Moscow would bring to bear all of the propaganda instrumentalities and arguments that it is now employing against alleged U.S. intentions to inaugurate a new arms race. It would target a great deal of attention to the militarization of space issues and to charges that the U.S. was violating the Space Treaty of 1967, the ABM Treaty of 1972, dozens of specific UN resolutions, the UN Charter, and so on. The fact that High Frontier systems are nonnuclear would not affect the propaganda effort. Moscow would disregard all facts in feeding its propaganda mill and would treat as truth anything it could dream up, as it is already doing with regard to the Space Shuttle.

This last point underscores a problem Moscow would face in its propaganda reaction: its ongoing propaganda campaign has all but exhausted the list of "crimes against humanity" charges that Moscow can voice against the U.S. It consequently might well be caught up in the classic "cry wolf" syndrome.

Politically, Moscow would try more of what it is already doing. All-out efforts would be made to promote the draft treaty barring any and all weapons in space that the USSR submitted to the UN in August 1981. It would intensify the "don't rock the boat" sentiments with regard to U.S. allies and would otherwise try in every way to generate pressures by the allies on the U.S. As for pressures on the U.S. itself, Moscow's initial reaction would encompass a mixture of carrots, in the form of apparent concessions in the arms negotiation area, and sticks, in the form of threats to push its own arms efforts in space and elsewhere to the absolute limits. There would be threats to repudiate the ABM Treaty, with a likely preference being to stigmatize the U.S. as killing the treaty or acting in what Moscow would interpret and proclaim as a blatant disregard of its terms. Should the initial carrot and stick mix fail to produce results, Moscow might well increase the carrot portion, perhaps to the extent of offering meaningful concessions in the arms control area, but almost certainly not to a point of significantly reducing advantages presently enjoyed by the USSR.

A second step Moscow would certainly take would be to reorder its own armaments efforts. Space capabilities would be moved into a position of top priority. As was stated by *Izvestiia* on August 20, 1981, the official Soviet posture is: "The Soviet Union has, of course, the necessary means and capabilities for confronting the space ambitions of the overseas strategists." Given the activities and successes that have marked the Soviet space program since the late 1960s, and particularly the progress that has been made toward establishing and using a system of manned space stations, the USSR probably does have a far greater military capability in space than is generally recognized. It is also a distinct possibility that Moscow would unveil some of these capabilities in the hope of intimidating U.S. leaders. However, Moscow has been talking about these capabilities in terms of the Space Shuttle. It is highly unlikely that existing capabilities encompass anything that would make it possible to negate or seriously affect High Frontier resources, short of acts of war.

It may well be that existing Soviet capabilities are adequate for the USSR to mount a High Frontier system of its own. This in itself would in no way interfere with or affect the U.S. system. The only effect would be to have antimissile systems on both sides, with each incapable of harming or affecting the other. In that case, the strategic situation of the USSR will have been profoundly altered. Instead of enjoying a one way system of deterrence based upon an illusory Mutual Assured Destruction concept, it would have to adjust to a two way system based upon a Mutual Assured Survival situation.

Such a change would wreak havoc with strategic elements of the Soviet war-fighting and war-winning edifice that Moscow has structured over the past 20 years. Moscow has truly operated in the war preparation sphere on the basis of the proposition that Khrushchev enunciated for the benefit of Kennedy at Vienna in June 1961: "Missiles are the god of war today." The tasks the USSR will face if High Frontier becomes a reality require high technology on a prodigious scale. The Soviet economy, already severely strained, may well be unable to meet these requirements for high technology without disintegrating.

A final possible Soviet reaction that must be considered is that the Kremlin might simply refuse to stand by while the U.S. puts its new High Frontier systems into operation. It might instead take advantage of the "window of opportunity" that stands open as a result of the USSR's present military advantages. It is indeed a hard fact of life that a window of opportunity is open to the Kremlin. That is why something like High Frontier is so obviously necessary if the U.S. is to be able to ensure its long term survival without subjecting itself to increasing subservience to the USSR. The issue now is whether the window looks big enough for the Kremlin to calculate that the USSR can move through it without dangerous consequences to itself. If so, a strong probability must be allowed that the Kremlin will use the window at any point in time it is provoked or opposed by the United States. If not, a strong probability must be allowed that the Kremlin will not risk its use even if faced with the certainty of complete success for High Frontier, for High Frontier will not in any way threaten the existence or even the well-being of the USSR. As a senior Soviet diplomat stated in the early 1970s, "ABMs don't kill people; ICBMs do." All it will do under the best of circumstances is to deny to the USSR the ability to threaten the existence of the U.S. Given this, all that is known about Soviet leadership indicates it is highly likely to choose to wait until another and surer day.

THE WEST EUROPEAN REACTION

For responsible West Europeans, U.S. adoption of the High Frontier strategy, if backed by convincing evidence of a U.S. will and ability to make it work, would lead to a revival of faith in the effectiveness and reliability of the U.S. deterrent and hence would be viewed as a godsend to their security interests.

From the West European standpoint, deterrence as a viable strategic concept has always depended upon whether it would safeguard West Europe from Soviet attack. As long as the U.S. had a monopoly or near monopoly of nuclear power, Europeans took for granted that this would be the case. Both the logic of the importance of Western Europe to the U.S. and the U.S. commitment to its defense through NATO confronted the USSR with the apparent certainty of nuclear devastation in case it resorted to overt aggression in Europe.

However, as the USSR developed mounting capabilities to inflict nuclear devastation on the continental U.S., skepticism has grown among Europeans that the U.S. would actually risk its own destruction in their defense. The Soviets have consistently fed this skepticism, contending that the U.S. would be deterred from using its nuclear power not only because it would be destroyed by a nuclear war but also because the capitalist system as a whole would be destroyed. Moreover, since the late 1950s, they have persistently argued that whatever the U.S. did, it would not help the Europeans, since the total destruction of any and all European countries that associated themselves with the U.S. in a war with the USSR would be the first and surest result of that war.

In European eyes, the U.S. movement toward a doctrine of Mutual Assured Destruction appeared to confirm their worst fears. If, in fact, the U.S. was prepared to accept that any use of nuclear weapons would constitute an act of suicide on its part, then it was simply inconceivable that the U.S. would employ these weapons in any circumstance short of a direct attack on its own territory. Many doubted it would do so even in that extreme.

In case of a Soviet advance into Western Europe, the U.S. stance appeared dangerously ambiguous to the Europeans. On the one hand, there remained the U.S. commitment to participate in the defense of Europe and to provide and employ for that purpose theater nuclear weapons, including those capable of reaching key targets in the USSR. On the other hand, there was doubt that the U.S. would respond to a Soviet attack by immediate employment of its full arsenal of strategic weapons. This seemed to mean that, at best, the U.S. would put its own territory at risk only after the West European countries had been devastated by theater nuclear exchanges. The worst fear was that the U.S. would simply hold back its strategic nuclear power in the hope of deterring the USSR from nuclear strikes against North America.

Thus deterrence for the Europeans has come to be a strategy of faith, rather than one of reasonably assured successful defense. Officially, West European governments have continued to adhere to it because they have no alternative. However, they increasingly give reasons for doubt that they would act out their role in the U.S.-conceived scenario of theater defense should the USSR decide to test that scenario. Current attitudes of the peoples of the West European countries justify an assumption that the governments would have real difficulties in offering resistance on the order planned even if they decided to do so. Careful analysis indicates these attitudes reflect pervasive popular feeling among West Europeans and are not simply a display of contrived anti-Americanism on the part of small groups of extremists.

As the current antinuclear leftist campaign in Western Europe grows steadily stronger, the political prospects for NATO Theater Nuclear Force modernization become less favorable. The incremental approach to improving the situation with additional nuclear weapons, as dictated by the U.S. strategic doctrine of deterrence, now seems to be counterproductive. A new answer to NATO military problems is becoming absolutely necessary.

A new U.S. doctrine and force posture which shifts emphasis from an exclusively offensive stance to a balanced offense-defense mix to make the use or threat of use of Soviet nuclear weapons ineffective and to assure mutual survival rather than mutual destruction would eventually appeal to the European allies. It would not, of course, turn the antinuclear, anti-U.S. element completely around. That group will quickly regear to blast the new strategy as provocative and likely to increase the probability of nuclear war.

A new, genuinely defensive strategy, carefully conceived and presented as one designed and phased to gradually strengthen deterrence, take the world out of the shadow of nuclear terror and incineration, and improve the security of the West against nuclear war, should command the support of responsible European political leaders, defense-knowledgeable elites, and many opinion leaders in Western Europe.

High Frontier would thus profoundly alter the strategic situation of Europeans. Its formal announcement, or its quiet inauguration, would not in the short term quell the popular turmoil in Europe. Indeed, the immediate effect might well be to increase turmoil because of fears that the USSR might be provoked into some sort of action, the brunt of which would fall first and foremost on Europe. Further, there would be strong reactions by many to the danger of extending the weapons contest into space. Some government leaders might urge delays in the hope of not rocking the boat at this particular juncture. Except for a hard core of pro-Soviet elements intermixed with the European equivalent of U.S. diehard believers in the efficacy of negotiated arms control, demonstration that the U.S. was determined in the course it had chosen would lead to the following decisive changes in European attitudes in a relatively short time:

- There would be a realization that the U.S. was beginning to break out of the paralytic bonds imposed by the concept of Mutual Assured Destruction.
- Even more decisive would be the realization that High Frontier would provide protection for Europe from Soviet launched strategic ballistic missiles. The absence of such protection, and the resultant forced and exclusive reliance on the deterring effect of threatened retaliation, have been the sources of the great nightmare with which the Europeans have had to live since the USSR began its deployment of a multitude of medium range missiles capable of reaching all points of European and adjacent territories. For the Europeans this would mean a strategic turnabout of momentous proportions.
- Finally, there would be a restoration of the badly shaken European confidence in U.S. ability and resolve to actually use its power to preserve the Free World.

CHAPTER VI ANNEX: ECONOMIC CONSIDERATIONS

INTRODUCTION

What would happen to the economy of the United States and of the Free World in the event of a major thrust into space by the United States beginning in 1982? Economic forecasting is not a rigorous, testable science like astronomy or physics; the answers to such a question are necessarily qualitative. Although market surveys and return-on-investment calculations can be made for specific activities which might be conducted in space in the near future, considerations of the overall economic structure of the U.S., historical analogies, and public psychology are probably more convincing and useful.

Space industrialization (including military activities in space) may be defined generically as the extension of human activities beyond the biospheres of Earth. These activities can be categorized in the following hierarchy:

PRIMARY ACTIVITIES

The extractive industries (hunting, fishing, agriculture, forestry, and mining) obtain the raw materials necessary for survival and for all other activities. Primary energy sources, such as fossil fuels, biomass, and uranium, are included here, as well as hydroelectricity.

SECONDARY ACTIVITIES

The processing industries (manufacturing, construction, petroleum refining, ore smelting, food processing and packing, and electrical generation from fossil and nuclear fuels) convert raw materials, or already processed materials, into more useful or versatile products or forms. Steel mills convert coal and iron ore into a variety of grades and compositions of steel; rolling mills convert bulk steel into beams, plates, sheets, and wires; automobile factories convert sheet metal into automobile fenders; etc.

TERTIARY ACTIVITIES

The service industries (transportation, communications, news media, basic education and job training, health care, insurance, banking, legal services, military defense, much of civil government, etc.) facilitate and support all other activities. Mail and telephone services support the railroads and airlines, banking supports telephone services, data processing services make banking possible. Without these service industries, all extractive and manufacturing industries would remain cottage industries.

QUARTERNARY ACTIVITIES

Activities carried out for their own sake or for the personal satisfactions they provide constitute this category. These include socializing, sports, hobbies, much of higher education, cultural activities, pure research, fine arts, music, vacation travel, etc. On the whole, the purpose of primary, secondary, and tertiary activities is to make possible the pursuit of quarternary activities.

Historically, the earliest space systems were deployed as purely scientific research projects in the interest of national prestige and, thus, were quarternary activities, carried out for their own sake and exempt from normal benefit-to-cost ratio analyses. During the 1960s, however, space technology was applied to weather observation and communications satellites, so that space industrialization came to encompass some tertiary activities as well. The focus both in funding and in visibility, however, remained on the Apollo program, a pure example of a quarternary activity.

As the expansion of satellite communications became more profitable in direct economic terms, launch services grew in importance justifying development and deployment of such projects as the Space Shuttle, the European Ariane vehicle, and the Japanese N-family of vehicles, in a further growth of tertiary (service industry) activities. In the near future, the Space Shuttle will be used for experiments and small scale production in zero-gravity materials processing and for on-orbit construction and assembly, beginning secondary economic activities in space.

Finally, serious attention is now being given to energy production or harvesting in orbit and to mining of nonterrestrial materials for use in space and later on Earth. When the first such system comes on line, the whole spectrum of economic activities, from primary through quarternary, will be conducted in space as they are on Earth.

The key factor in this historical pattern has been the cost of access to and from orbit. With the Space Shuttle, the development of the transportation infrastructure has made a major leap forward, comparable in importance to the Erie Canal and to the Transcontinental Railroad during the 19th century. Next generation launch vehicles presently on the drawing boards can lower the cost-per-pound of access to orbit another 20 to 50 fold. This will permit the development of yet another quarternary economic activity on the High Frontier—namely, space tourism.

If space industrialization is merely the extension of economic activities into a new location, beginning just 100 kilometers overhead, why bother giving it any particular attention? Is there anything different or unique about it, justifying the apparent higher costs of doing business in space?

In economic terms, space is inherently different because of its unique attributes: weightlessness, abundant supply of cheap and clean energy, hard vacuum, easy access to a wide range of thermal conditions, excellent visibility from above, limitless room, isolation from the biosphere, and vast mineral resources. These factors taken singly or in various combinations imply that, in contrast to Earthbound industries, industrial facilities in orbit offer the potential for (1) higher levels of productivity, (2) rapid (even exponential) growth, and (3) the creation of new wealth from completely new resources.

Several studies in recent years have been performed on the economic potentials of various industries in space. Perhaps the most systematic of these studies (at least of the civilian prospects in space) are the parallel studies on space industrialization over the period 1980 to 2010. These were performed under contract to NASA's Marshall Space Flight Center by Science Applications, Inc. (SAI), and by Rockwell International. The SAI study developed a useful scheme for characterizing space industries, which were catalogued in three groups:

INFORMATION

This included remote sensing of weather, Earth resources, ocean surveys, communications, navigation and positioning, scientific space exploration, and data processing and storage in space.

MATERIALS

Zero-gravity processing of materials brought up from the ground represents the earliest activities in this area. Mining of nonterrestrial materials and smelting of asteroid ores for use in space and on Earth are later possibilities.

ENERGY

Harvesting solar energy in orbit for delivery to the Earth by microwave or laser transmission or by direct reflection of sunlight are the main possibilities here. Other concepts include relaying power between continents via satellites and disposal or storage of nuclear wastes in deep space.

To these areas of civilian space industrial activities, we must add the following three areas of military activities:

FORCE DELIVERY

This includes antisatellite systems, laser battle stations, tactical projectile weapons, and global ballistic missile defense systems.

COMMAND, CONTROL, COMMUNICATIONS, AND INTELLIGENCE (C^3I)

This area overlaps with the information category, including military communications and weather satellites, but adds surveillance and reconnaissance satellites, theater-wide navigation and communications from orbit, orbiting command posts, and launch on warning systems.

MILITARY SUPPORT

A number of areas mentioned above can provide support to force delivery and C^3 operations, but some of these applications must be singled out for specific consideration here. These include repair on orbit, upgrade on orbit, propellant and consumables supply facilities, and housing and medical facilities in space. Nonterrestrial mining of strategic materials may also be considered a support service.

To the extent that development and deployment of military space systems assists the development of civilian facilities or stimulates technological innovations in productive systems, or to the extent that space based military systems are less expensive than ground based systems designed to serve the same goals, the real productive capacity of the nation will be increased.

In the following sections the economic effects of a High Frontier program are addressed from several different viewpoints using, in varying degrees, the framework presented above.

DIRECT BENEFITS

While hundreds of potential products and services which could be provided by industries in orbit have been described in varying degrees of technical detail, only a few dozen have been investigated for potential market size, revenues, and return on investment. The aforementioned SAI study evaluated several promising information services, material processing applications, and people-in-space activities, projecting revenues for these over the period 1980 to 2010.

The table below lists the evaluated services and products. Previous studies of the market potential for energy from space (SPS only) were used, together with a new market analysis for night illumination satellites. The SAI cumulative projection for specific products and services is shown below. Because this projection includes only some of the possiblities, it is likely to be conservative. (On the other hand, the particular products and services included may be poor choices that may not succeed for unforeseen reasons.)

Services and Products Included in the SAI Market Analysis

Information

Portable (wrist) telephone	3-D holographic teleconferencing
National information services	Direct TV broadcast (U.S.)
Disaster communications	Vehicle inspection
Global search and rescue	Nuclear fuel locator
Air traffic control services	Rail anticollision system

Personal navigation sets	Vehicle/package locator
Voting/polling sets	Education by TV (U.S.)
Urban/police wrist radio	Coastal anticollision system
Land and water resources	Ocean resources
Electronic mail	Power network monitoring
Materials Processing	
Drugs and pharmaceuticals	Semiconductor materials
Superconditioning materials	High strength magnets
Fiber optics	Perishable cutting tools
Bearing materials	Jewelry
People in Space	
Space tourism	Space hotel
Entertainment	Movies
Sports	
Energy	
Solar power satellites	Night illumination

Several key points emerge from this market analysis. First, total revenues from industries in orbit can grow to significant levels in comparison with the present U.S. Gross National Product. These revenues may increase to one or two percent of present GNP by the turn of the century and to four to six percent of present GNP by 2010. The SAI study was carried out in 1977-78 and already appears hopelessly conservative. More recent studies and assessments of just the communications services now suggest 20 to 40 percent growth rates through the 1980s for satellite communications, with revenues by 1990 in the range of $50-100 billion annually, nearly five percent of present GNP.

Second, information services will remain—as they have since 1965—the largest profit center in space for the next few decades, although the mass in orbit and the number of workers in orbit will be modest.

Third, energy from space has a huge potential for growth, with growth rates in the next century estimated higher than information services. (The market analysis includes *only* the U.S. market for SPS. India alone, by the turn of the century, could absorb the power output of 75 SPS units of five gigawatts output each, if they could be built that fast.) Total mass in orbit and workers in orbit would be immense if SPS is part of the total program.

Fourth, reflecting the structure of the existing economy today, the materials industries will be much smaller in total revenues than either the information services or energy production.

Clearly, the potential is huge. We are today in a position analogous to the investors in the Virginia Company early in the 17th century. The Virginia Company was chartered by the Crown of England as a profitmaking, joint-stock company. The founders of the company had a lengthy list of schemes for making money in the New World. These schemes, based largely on rumors and wishful thinking about the New World, included visions of abundant gold waiting to be picked up or seized from ignorant savages. None of these schemes actually worked, but the Virginia Company was an economic success because of an unexpected windfall. Once the colonists established a toehold on the new continent, they discovered that tobacco could be raised commercially for export to Europe. Doubtless many of the concepts considered today for commercial use of space will prove in hindsight to have been far-fetched. Nonetheless, we know far more today about the nature of outer space than the original shareholders in the Virginia Company knew about the New World. We too will encounter surprising and highly profitable uses of space surpassing our present imagination, if we only begin to establish a beachhead on the High Frontier.

TECHNOLOGICAL INNOVATION AND ECONOMIC GROWTH

The U.S. economy has been in difficulty for nearly a decade. A variety of causes has been postulated, ranging from changes in demographic structure to unfair competition from abroad and overregulation of industry. While most postulated reasons have some validity and a number have been addressed by attempted remedies in previous administrations, little more than lip service has been given to loss of industrial productivity due to declining innovation in the technological base of our industries.

The economic growth of the United States during the period from about World War I until the OPEC embargo has depended almost entirely on technological innovations made in the latter part of the 19th century. Industries which today form the bulk of the economy and are rooted in that age include steel, electric utilities, automobiles, aviation, radio and television communications, petroleum, and chemicals. During the economic growth of this century, progress has been essentially continuous, relying on incremental improvements in these basic technologies rather than on fundamental, qualitative breakthroughs.

Peter F. Drucker describes the present situation as an "age of discontinuity" (*The Age of Discontinuity: Guidelines to Our Changing Society,* Harper & Row, 1968, 1969):

> Genuinely new technologies are upon us. They are almost certain to create new major industries and brand-new major businesses. The growth industries of the (past) half century derived from the scientific discoveries of the middle and late 19th century. The growth industries of the last decades of the 20th century are likely to emerge from the knowledge discoveries of the first 50 or 60 years of this century: quantum physics, the understanding of atomic and molecular structure, biochemistry, psychology, symbolic logic. *The coming decades in technology are more likely to resemble the closing decades of the last century, in which a major industry based on new technology surfaced every few years, than they will resemble the technological and industrial continuity of the past 50 years.* (Emphasis added.)

By consciously and deliberately targeting investments into key areas of technology to take advantage of new opportunities in the marketplace, including those areas of industry in which real productivity is halted or declining, a cycle of major economic growth can be initiated. Of the industrialized Free World nations, only Sweden has done this systematically, which resulted in the transformation of Sweden from a primarily agrarian society at the end of World War II into a highly industrialized nation. Major export industries include automobiles, where Volvo and Saab pioneered the combination of consumer styling, low gas consumption, and mechanical durability, and airplanes, where Saab and Viggen pioneered short-field fighter aircraft.

The key question to address in the context of "reindustrializing America," then, is the following:

> What new technologies offer the greatest potential for significant economic growth in which existing knowledge and techniques give the United States a commanding lead over potential competition from abroad?

Several answers spring to mind immediately: computers, genetic engineering, and space technologies. The first two have already taken off, largely in the private sector, since the front end capital investments are comparatively quite modest. Pocket calculators now cost less than slide rules used to cost and are far more versatile, increasing the productivity of those who once used slide rules. (Slide rules have not been manufactured in the Western industrialized countries for about seven years because they have been displaced by pocket calculators.) In addition, a whole new industry, video games, has capitalized on this technology, with total revenues of more than $1 billion annually. The potential impacts of new computer technology are only beginning to be felt throughout the economy. Genetic engineering is still at a very early stage, but its impact on such industries as food production and on costs of pharmaceutical products and industrial chemicals will be enormous.

Space technologies, on the other hand, have not yet reached the takeoff point. This is due in large part to the high costs of transportation to orbit. Other factors include investor uncertainties about markets, perception of extended delay before an investment return, perceived and real regulatory inhibitions, and industry's perception that U.S. foreign policy has not reached a posture of clear support for private enterprise activities in space in the face of potential opposition and hostility from abroad. Most of these disincentives can be alleviated at no direct government expenditure, with the simple decision to make a national commitment to the High Frontier.

A coordinated High Frontier program in which civilian and military space systems are encouraged to share would provide the focus for targeted development of this major new technology. Technological development should be guided by specific marketable products or services to maximize economic benefits. Development of large space structures, for example, should lead deliberately to applications such as direct TV broadcast satellites and solar power satellites. Development of life support systems for space should be planned for materials processing facilities, on-orbit repair facilities, and survivable command posts in orbit. As has been the case with the microelectronics revolution, these technological innovations would find countless unforeseen applications elsewhere in the economy, with positive effects on productivity and thus on economic growth.

Would the investments needed for this program be better spent in some other area? Certainly this is a debatable point, but several considerations strongly favor space technologies. In the short run, most of the investments will be channeled into the existing aerospace industry. Of all U.S. industries, only retail sales employs more people per dollar of capital. Expanding the size of the aerospace industry would create more direct jobs than virtually any other allocation of effort. Previous studies (Chase Econometrics, for example) have shown very high "multipliers" for aerospace investments as well. Each dollar spent on the space program during the 1960s, for example, generated $6 or $7 of new GNP over the ensuing few years, supporting other jobs in the economy as well.

In the long run, no other technology in our grasp offers access to such huge new resources or to the prospect of creating totally new wealth from sources outside the present economic system. An investment during the next 5 to 20 years in space can provide access to the entire solar system—not just for the United States but for all the world. If the U.S. does not take the lead, those resources may well be developed by others, notably the USSR, with far less willingness to share access.

We should not allow energy price competition in the United States to blind us to the attractiveness of exporting energy from solar power satellites to the underdeveloped world. Since the entire economic infrastructure is very inadequate in the less developed countries, including especially the transportation industries, costs of energy are typically three to five times higher than in the

United States. The rise in OPEC petroleum prices has aggravated the energy crisis faced by these nations. Building nuclear power reactors in Third World countries is much more expensive than building the same reactor in the United States because many of the parts, which are inherently high technology products, must be imported through inadequate transportation systems. Most of the construction workers must be highly skilled and thus must also be imported. The savings on energy costs which might be expected on the basis of the difference between nuclear-generated electricity in the U.S. and alternative energy systems in the Third World are severely compromised by higher installation costs, as well as political difficulties.

In the case of solar power satellites, however, more than half the cost of each generating plant is attributable to the space segment. No cost differential applies between a highly industrialized country and a developing country. The receiver antenna system (rectenna) consists of very few high technology components, viz., the solid state rectifier elements. These could be manufactured with highly automated equipment in urban factories in developing countries, with limited numbers of skilled workers. The overwhelming bulk of components for the rectennas could be assembled by semiskilled workers, and construction of the rectenna could be done on-site by large numbers of unskilled workers with semiskilled supervision. Thus the capital costs for a complete SPS would be only slightly higher in a developing country than in the United States. Energy costs would then be substantially lower than present sources such as petroleum, firewood, or cow dung.

MILITARY CONSIDERATIONS

A national commitment to the High Frontier program would harness technology developments aimed at military purposes in support of economic productivity. Economic benefits are certain but difficult to estimate in quantitative terms. More importantly, if the strategic purposes of the United States can be achieved as well (if not better) by going into space rather than by relying on more conventional ground based systems, then the Federal budget can be reduced by the difference in system costs. This in turn would reduce government borrowing during the next few years, easing pressure on interest rates. The effects on the overall U.S. economy of lowered interest rates are discussed below. More specifically, however, if reduced Federal demand for borrowing eases interest rates, the cost to the Federal government of servicing the national debt, which is nearly $1 trillion now, would be reduced by $10 billion annually for each percentage point decline in interest rates.

ECONOMIC EFFECTS OF RENEWED HOPE

With all the attention devoted in recent years to perceived scarcities of clean air, clear water, petroleum, natural gas, strategic materials, investment capital, and government funds, we have barely noticed that the most critically scarce commodity of all has been "hope." In the growing expectation that things are only going to get worse, consumers have had little, if any, incentive to save. Consumer debt has been climbing at ever faster rates. Unemployment among teenagers has convinced much of American youth that the future is bleak, contributing to a major crime wave and to increased drug usage, with very large economic costs to all of society. The only segment of the public which is swimming upstream and struggling to get completely out of debt is the survivalist movement, people who are becoming convinced that a major economic and possibly social collapse is imminent.

A commitment by the United States to strong use of the High Frontier could dramatically change public attitudes and instill a new sense of purpose and hope in this nation and in the rest of the Free World. A major thrust into space would provide the world with clear and convincing evidence that the resources available to the human race are not fixed and that new wealth can be created without depriving others. With renewed confidence, consumers would be motivated to save for and invest in the future. The prospect of new jobs in space industries and even of jobs for skilled hardhat workers in orbit can be expected to improve the morale of unemployed youth and renew their faded dreams.

PUBLIC PSYCHOLOGY AND THE ECONOMY

In this discussion, much emphasis has been placed on the economic effects of changes in public psychology. The role of these psychological factors in economics can be highlighted by the following anecdote. Several years ago, Eric Burgess, one of the founders of the British Interplanetary Society, was discussing some very ambitious future space project with a successful and very wealthy financier. When the financier asked how much the project would cost, Burgess apologetically quoted a huge number and said, "Unfortunately, it's just too expensive to afford." The financier immediately retorted, "Nonsense! We *invented* money, didn't we?"

A national commitment to the High Frontier, if carefully presented to the American public and to the Free World community, could quickly alter public psychology, especially in its views of the future. Such changes would alter economic realities for the better more rapidly than any amount of tinkering with social programs or Federal Reserve discount rates. Some of these effects are discussed below.

PROMPT ECONOMIC EFFECTS

Two important short term economic effects can be expected to result from a major national commitment to the High Frontier program: (1) increased savings and investments and (2) a softening of world energy prices, especially for petroleum. These effects would arise from changes in public psychology rather than from shifts in cash flows in the economy or from actual growth in productivity. Since changes in public psychology can take effect within as little as a year, major efforts should be made to display a high degree of national commitment to the program to guarantee that these effects materialize.

The going interest rate is determined by the anticipated inflation rate and by the average discount rate. If either rate can be lowered, the interest rate will inevitably decrease as well. When a potential lender or investor has disposable funds available, he can either spend them immediately to obtain personal gratifications (e.g., trips, new clothing, restaurants, luxury gifts, newer or more luxurious car or home, etc.) or he can invest the money, deferring that spending until a later date. An inducement must be offered to the lender in exchange for the deferral of gratifications. That inducement is the discount rate, and each potential lender has a different discount rate. When the demand for loans rises, a larger fraction of those lenders whose personal discount rate is high become willing to lend their money rather than spend it on immediate gratifications. The average discount rate is then the rate at which the supply of investment capital and the demand for investment capital balance to clear the market.

The expectation of technological innovations resulting in superior products or services at the same cost or the expectation of increased productivity lowers the discount rate for all investors. The result is that more money is saved or invested. More investment capital becomes available for reindustrialization, while interest rates remain the same or even decline. The availability of more capital for plant expansion or for replacement of obsolete, less productive equipment would improve productivity throughout the economy, further stimulating higher expectations and further lowering personal discount rates for potential investors. The expectation of technological advances and the resultant incentive to capital investment would almost certainly ensue from a vigorous commitment to the High Frontier.

If the High Frontier program were to include a strong commitment to power from space on economically significant scales, OPEC petroleum prices would soften significantly within a few years. Petroleum reserves in the OPEC nations are large, but finite. Prices and production levels are set to maximize the total accumulation of capital from petroleum sales over the few decades before the cheap supplies run out. If it becomes clear to petroleum producing countries that the demand for petroleum will decline significantly in 10 years or so, the value of oil retained in the ground would decrease, production levels would increase, and prices would stabilize or decrease.

This effect, even more than the improved investment market, would depend on a convincing and credible U.S. commitment to a major new energy source which promises to be economically competitive.

The benefits of a softening or decline in world energy prices are fairly obvious. Less money spent for each unit of energy means that more energy can be obtained and used for the same outlay of funds or that money is freed up for other, more constructive purposes. In the industrialized nations of the Free World, energy consumption would likely rise slightly in response to the price changes, with most of the change in funds becoming available for other uses. In the less developed countries, consumption of petroleum would accelerate, since scarcity of funds to pay for petroleum imports has severely constrained use of petroleum fuels. This would strengthen the industrial base in the developing countries with major benefits to both agriculture and the environment. This is because the scarcity of fossil fuels during the last decade has resulted in accelerated deforestation and in the use of animal dung for fuel rather than for fertilizer. Major financial institutions in the United States have lent massive sums to developing countries to finance their petroleum imports. Billions of dollars of these loans are in risk of default, even with low interest rates subsidized by development agencies such as the World Bank. The softening or decline of petroleum prices, by the strengthening economies of developing countries, would lessen fears in the financial community of defaults on these international loans.

IMPACTS ON JOBS AND TAXES

The SAI study forecast some jobs created and new tax revenues generated for the Federal government under three different assumptions about the space industrialization programs over the next three decades. The three scenarios examined were:

- Baseline case, with no SPS program.
- Baseline case, including SPS.
- Upside program, in which foreign competition and some incentives to private enterprise result in a vigorous civilian program.

Both baseline cases, just four years later, already appear hopelessly conservative due to explosive growth in commitments for communications satellites. The High Frontier program envisioned here would encompass a higher level of national commitment (due to the inclusion of the military space systems) than was envisioned in the Upside program. (The SAI study was conducted under ground rules which specifically forbid consideration of military activities in space, let alone synergisms between civilian and the military space efforts.) Given all these caveats, the table below shows these conservative projections.

New Jobs and Taxes Generated

New jobs (direct only, based only on U.S. markets for space industries.):

Year	No SPS	Baseline, with SPS	Upside
1985	15,000	100,000	120,000
2010	1,000,000	1,900,000	3,800,000

Taxes generated (direct only, based only on U.S. markets for space industries, 1977 dollars):

Year	No SPS	Baseline, with SPS	Upside
1985	$ 100 M	$ 800 M	$ 1,000 M
2010	$10,000 M	$20,000 M	$40,000 M

The SAI report comments on these results as follows:

> The estimate of jobs for 1985 is probably low by a factor of two since most funding would be to the aerospace industries. The Aerospace Industries Association (AIA) has estimated that about 30 direct jobs are created for each one million dollars of appropriation. Direct plus indirect jobs are estimated to total about 100 jobs per one million dollars. Thus, the job projection for 1985 is conservative . . . . The true impact on new jobs is some two to four times the numbers shown here, depending on specific assumptions and economic theories applied.
>
> Exactly how many of the new jobs are displacements of old jobs or creation of new ones is difficult to speculate. Very little displacement is anticipated since most of the new capabilities afforded by space industrialization are complementary to existing systems. . . .
>
> In the aggregate, the best guess is that 75 percent or more of the postulated space industrialization initiatives revenues will be job creating in the 1990s and beyond. Thus for a [total U.S.] work force of 100,000,000 in 2010, some three to twelve percent could be employed in new jobs created by space industrialization.

CHAPTER VII: IMPLEMENTATION

To achieve the economic and national security objectives set forth in High Frontier, several new civil and military space and supporting systems need to be developed as expeditiously as possible. Not only must the initial first generation systems be acquired quickly and economically but they must be acquired in a way which sustains their priority nature and avoids any sizable near term budget add-ons or fiscal year "balloons."

The time needed to implement the proposed space capabilities, especially on the military side, is critical to the effort's overall ability to recover the margin of safety that the President has promised. It is also crucial to ensure that the Soviet Union does not achieve the capability to deny the U.S. access to space for either national security or economic purposes.

THE PROBLEM

The average of 13 or more years it now takes DOD to acquire major new weapons systems is unacceptably long and enormously expensive. In the 1950s, strategic systems such as Atlas and Polaris were selected, developed, and made operational in four to six years. At the time of their development, these systems had more technological unknowns than do many of the illustrative systems discussed in this report.

Of the numerous studies during the past 10 years on how best to acquire new defense systems, two have had recent impact: (1) the Defense Science Board 1977 Summer Study of the Acquisition Cycle, chaired by Dr. Richard D. DeLauer, now Undersecretary of Defense for Research and Engineering; and (2) the April 10, 1981 "Memorandum on Improving the Acquisition Process" by Deputy Secretary of Defense Frank Carlucci. Both reports are in general agreement as to why the time for new defense system acquisition has more than doubled since the 1950s. They also agree on corrective measures required.

NASA does not suffer as much as Defense from such things as the overlayered organizations, the competition of multiple systems for funding, and the many regulations that now stretch out acquisition times. Nevertheless, special implementation measures will be required to expedite the NASA-managed systems as well as those managed by DOD.

The factors introduced over the past two decades most responsible for the overly long time it now takes to acquire advanced new weapon or space systems are:

Front end decisionmaking. This is time devoted to stating a firm operational requirement and selecting specific systems for full scale development. The decisions necessary to implement the High Frontier concepts can and should be made by the President on the advice and recommendation of a committee of recognized scientists and systems acquisition experts.

Inadequate commitment to acquisition from the outset. Funding and assigning top priorities for full system acquisition are necessary to preclude delays in the program resulting from having to compete for or await further funding in latter phases. This can be prevented in the case of first generation High Frontier systems by centralized advocacy for multiyear funding and "fencing" of the appropriations obtained.

The reticence in recent years to maximize concurrency in the acquisition process. Failure to initiate procurement of long lead time production items during the development and testing phase; to authorize full production prior to "first flight"; and to conduct joint technical, operational, and user evaluation testing has added years to achieving operational capabilities.

Insistence on too detailed and specific performance requirements or operational capabilities. The first generation of new, "cutting-edge" space and point defense systems essential to the implementation of this proposal can benefit from cost and time saving contractor-proposed innovations, provided that the performance and operational capabilities specified are not too rigid. Departments must write broad, flexible specifications stating only what is needed and require only "sufficient" performance for the first generation High Frontier systems. Industry should also be allowed as much flexibility as possible in responding to requests for proposals.

Overregulation of the acquisition process and excessive organizational layers in the review, approval, and decision process. Timely procurement and cost savings will require waivers from strict adherence to the acquisition and procurement regulations of the Office of Management and Budget and DOD. These should only be used as guidelines. The chain of command for review and decisions for High Frontier systems in all departments should bypass all levels not in a position to make authoritative decisions on the issues concerned. A streamlined decision, review, and procurement channel, with High Frontier expediting offices at appropriate levels, should be set up to avoid delays.

Personnel selection and motivation. Unless experienced, dedicated program managers and key people are assigned to the task at all levels and remain with the program throughout the acquisition cycle (four to seven years), time and effectiveness will be lost. Departments should take appropriate steps to meet these requirements, including provisions for promotion and protection from any loss of career opportunity. Similar provisions should be made for key personnel in the High Frontier centralized organizations.

Distrust and misuse of contractors. The complexity of advanced new systems, the frequent need for decisions as to changes or alternative approaches to resolve technical problems, and the ability to exploit money saving contractor suggestions require close teamwork between the contractors, the program managers, and the eventual user organizations. In recent years, a distrust of contractor motivations and/or fear of adverse publicity and criticism resulting from close personal relationships between government officials and contractors have worked to the detriment of the team effort that is absolutely essential to the efficient and economical acquisition.

Excessive micromanagement. Appropriate decisions should be made at those levels where the expertise exists, to prevent wasted time and costly mistakes. In recent years the growth of congressional staffs has all too often resulted in young, dedicated but inexperienced congressional staffers seeking to influence technical and operational decisions that should not be dealt with at the congressional level. The tendency towards micromanagement of systems acquisition programs by congressional or DOD staffers and principals at levels not directly responsible has resulted in delays, waste, and the risk of bad decisions.

BASIC IMPLEMENTATION CONCEPT

The basic implementing plan for first generation High Frontier systems is to provide for special, centralized policy and management only as deemed necessary to: (1) select the first generation systems to be acquired; (2) maintain the identity and priority of the "initiative"; (3) deal at the highest level with the funding justification and allocation and with the overregulation problems; and (4) ensure quick decisionmaking and risk taking where appropriate.

The acquisition task for each specific system

would be assigned to the responsible department presumably, but not exclusively, NASA and Defense, by Executive Order. This order should include Presidential instructions to the departments to take organizational and procedural measures necessary to minimize or eliminate delaying factors and achieve the desired operational capabilities at the earliest possible dates.

PROPOSED CENTRALIZED ORGANIZATION

The centralized Executive level organization would consist of three separate entities (see Figure 14). These would be established by the Executive Order and would exist only as long as necessary to acquire and deploy the first generation of new High Frontier systems. Housekeeping should be assigned to NASA. These entities would be:

1. A National Space Council with the Vice President as its chairman. Membership would be from the White House, NSC, DOD, NASA, DOE, Congress, and others designated by the President. The chairman would be the chief executive officer for the High Frontier initiative. This council would be similar in composition and mission to the National Space Council appointed in 1958 by President Eisenhower.
2. A Systems Selection Task Force similar to the "Von Neumann" Committee appointed in the 1950s which selected and recommended to the President the first U.S. ICBMs to be developed. Membership should consist of leading scientists, industrialists, strategists, and systems acquisition authorities appointed by the President. Its task would be solely to review industry and departmental proposals and recommend to the President the first generation High Frontier systems in order of priority that responsible departments would be directed to develop and acquire. A deadline of 90 days should be set for this task. The systems selection task force should have a functional staff and be authorized to task government, industry, or Federal contract research groups to accomplish its mission.

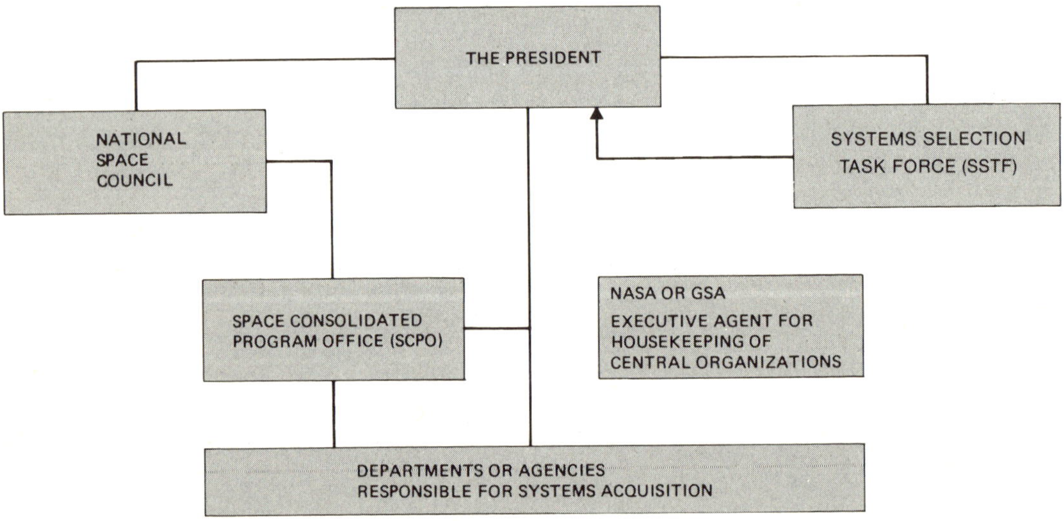

Figure 14. High Frontier Recommended Implementing Management Organization

3. A Space Consolidated Program Office to be headed by a program director who would be the chief operating officer for the High Frontier initiative. Consideration should be given to making the deputy administrator of NASA the space consolidated program director as an additional duty.

For the recommended terms of reference of these bodies see Appendix H.

DEPARTMENTAL ROLES

These three temporary, centralized organizations are essential to expedite acquisition of the first generation systems, sustain priorities and oversee implementation of the initiatives for improving the systems acquisition process. The government resources, as well as administrative and research support required to successfully acquire the new systems, are now vested in the user agencies. Duplicating these or transferring them to new centralized organizations would be time consuming and costly. It undoubtedly would also be strongly resisted, and any new organization could suffer from a reticence to serve on the part of many experts whose careers are tied to their departments or military services. Major legislative problems could also be encountered in connection with authorization, funding, and manning of any new separate organizations whose missions would be perceived to overlap those of existing departments.

In view of these considerations, the responsibility for the acquisition of each specific system should be assigned by Executive Order to the department having primary interest. The Executive Order should contain the special procedures and organizational relationships specified in Appendix H, Tab A to eliminate major delays in the acquisition process.

Presidential instructions to adopt some or all of these measures should not be deemed an infringement on departmental responsibilities. Bypassing established organizations such as the Joint Chiefs of Staff or military services can be done by Presidential authority. Any such instructions would routinely be reviewed by the proposed National Space Council where affected departments and agencies would be represented. Insofar as DOD is concerned, the proposed special management measures are supported by the findings and recommendation of the DOD acquisition studies, such as the DeLauer and Carlucci reports, and current DOD policy documents on systems acquisition (DODD 5000.1 and DODDI 5000.2).

CHAPTER VIII: TREATY CONSIDERATIONS

GENERAL

The High Frontier concepts and programs cannot be implemented without an impact on arms control negotiations, past, present, and future. At the core of High Frontier is a fundamental change from Mutual Assured Destruction (MAD) toward Assured Survival. We cannot make this strategic change without also rethinking our approach to arms control.

U.S. arms control efforts to date have been rooted in two basic precepts of MAD: stability depends upon a balance of terror to be sustained by a negotiated equality (or parity, equivalence, sufficiency, etc.) in *punitive* nuclear weapons and the inescapable corollary of that doctrine, strategic *defensive* weaponry, is destabilizing and provocative.

The MAD-based approach to arms control negotiations was articulated clearly by Secretary of Defense Robert S. McNamara in the mid-1960s. He stated:

> We think it is in our interest, and theirs, to limit the deployment of defensive weapons, and we're quite prepared to discuss possible limitations in the deployment of offensive strategic nuclear weapons as well. (Pentagon News Conference, Washington, D.C., May 18, 1967.)

Mr. McNamara calculated that the Soviets would be unlikely to embrace balance of terror as a basis of arms control negotiations as long as the United States maintained a superior position in strategic nuclear offensive power. He asserted that strategic stability and the conditions for effective SALT negotiations would be improved if the Soviets were allowed to increase their nuclear attack capabilities to a level where they would be certain of inflicting intolerable destruction on the United States in a retaliatory strike. He moved forcefully within the Department of Defense to derail strategic programs, defensive or offensive, which could thwart the achievement of this presumably desirable balance of terror.

THE INFLUENCE OF MAD ON SALT

The results of SALT negotiations thus far quite clearly demonstrate the effects of these MAD precepts. The only *treaty* resulting from SALT is the ABM Treaty which is designed to outlaw strategic defense, at least as far as defense against ballistic missiles, the most threatening of offensive systems. On the other hand, our attempts to negotiate limits on offensive systems, the Interim Agreement of 1972, the Vladivostok Accords, and SALT II, resulted in the ratcheting upward of the level of offensive nuclear weapons. Americans were urged to accept this ''progress'' in SALT on the basis of the MAD theory that nuclear war would be so apocalyptically destructive that its deterrence is independent of the numbers of weapons involved.

The U.S. has entered negotiations on offensive systems attempting to fix limits consistent with MAD theory—that is, at or below existing U.S. inventories and hoping to avoid any increase. The Soviets, on the other hand, enter negotiations determined to fix levels high enough to accommodate an entirely different strategy, which insists that nuclear war would destroy capitalist nations but that the socialist camp—despite widespread destruction—would emerge triumphant. The Soviet SALT negotiators insist that levels be high enough to encompass their ongoing weapons programs designed to support that strategy. Invariably, both sides are accommodated. The Soviets are allowed to pursue the war-winning capabilities consistent with their doctrines, while the United States is permitted to add to its retaliatory-only capabilities consistent with MAD theory.

The inevitable effect of negotiations based on these two fundamentally divergent strategic views has been an intolerable growth of Soviet nuclear first strike capabilities and a dangerous weakening of the U.S. deterrent. Western arms control advocates have been unwilling to accept the obvious reality that the Soviet Union rejects, in word and in deed, the MAD doctrine which underpins Western devotion to the SALT process. The Soviets have from the inception of the assured destruction theory branded it as "bourgeois näiveté." They have not elected to leave their homeland defenseless against U.S. nuclear retaliation, as MAD theory demands, but have poured more resources into strategic defenses, active and civil, than the U.S. has invested in its entire deterrent force. They have created offensive systems obviously designed to destroy as much as possible of the U.S. retaliatory force in a first strike.

This incontrovertible evidence of the Soviet strategic perspective and its incompatibility with the U.S. approach to SALT negotiations is swept aside by many arms control advocates by a far-fetched assumption that there are Western-style "hawk" and "dove" factions in the Kremlin. According to this assumption, Brezhnev and other "civilians" really do accept MAD theory but are opposed by a powerful group of Soviet "militarists" who insist that nuclear war is not only thinkable but winnable.

To accept this view of the Soviet leadership, one must make himself believe that the omnipotent Communist Party, headed by Marshal of the Soviet Union, Leonid Brezhnev, cannot control its comrades in the Red Army and indeed must reluctantly imperil the entire economy of the USSR to meet the demands of a Russian "military-industrial complex." If one can bring himself to believe this, he can then argue, as many arms control advocates do, that the U.S. should accommodate to intransigent Soviet positions in arms control negotiations in order to strengthen the hand of Kremlin "doves" in their difficult task of restraining the military "hawks." It would also follow that the key alliances in the arms control process are, on the one hand, between peace loving adherents of MAD theories in the U.S. and USSR and, on the other hand, between the "militarists" in the Pentagon and their counterparts in the Soviet Ministry of Defense.

As long as the MAD theory remains the basis for the U.S. approach to arms control, the SALT process will continue to undermine the security of the Free World. No SALT agreement agreed upon by U.S. negotiators attempting to establish and maintain a balance of terror and Soviet negotiators determined to establish and maintain strategic nuclear dominance will ever be ratified by the Senate of the United States.

Further, the longer MAD theory and arms control advocacy remain interlocking concepts among Western intellectuals, the more contrived will be the excuses provided for Soviet behavior whether in the SALT process or elsewhere.

CURRENT TREATIES

The High Frontier strategy of Assured Survival can be adopted and pursued without regard for further arms control agreements with the Soviets. Indeed, one of the salient advantages of High Frontier is that it provides security to the West quite independently of any trust or distrust of the leaders of the Soviet Union. The usefulness of High Frontier's spaceborne strategic defenses are not affected by Soviet compliance with past arms control agreements. This important advantage should not be affected by any future arms control agreements.

This is not to say that the High Frontier strategy excludes all consideration of arms control. In fact, the reemphasis of strategic defenses central to the High Frontier approach has received support from an unexpected quarter. Mr. Anders Boserup, a Danish activist in the international disarmament movement, states:

> . . . the adoption by states of a defense approach to security need not lead to an arms race. On the contrary, it can lead to disarmament, and is probably the only viable approach to it. (*Bulletin of the Atomic Scientists,* December 1981).

It would be naive indeed to predict that Mr. Boserup and his colleagues in the disarmament lobby would applaud U.S. adoption of the High Frontier strategy. They are much more likely to condemn it as the initiation of a "new arms race in space." Nevertheless, what he wrote is *true.*

Ironically perhaps, the employment of effective spaceborne defenses will accomplish, through unilateral U.S. action, that same result which the disarmers have so fruitlessly pursued over 15 years of SALT talks—the checking of the growth of nuclear offensive weapon inventories on both sides. Effective strategic defenses can negate the paramountcy of the nuclear ballistic missile in the strategic equation and eliminate the imperative on both sides to have more weapons with even greater destructive power. The U.S.-USSR competition would be shifted from a numerical contest in nuclear offense to a technical contest in defensive systems in space where nonnuclear technologies show great promise.

Even if nuclear weapons come to play a role in the defensive competition in space, the threat of their use, hundreds of miles above the Earth, would certainly be preferable to the threat they now pose in the form of ballistic missile warheads aimed at terrestrial targets.

The adoption of the High Frontier strategy, despite these advantages for the real world of peace and security from nuclear devastation, will require a fundamental change in the U.S. approach to arms control negotiations, which is certain to engender controversy. MAD theories will not die easily, in or out of government. There is no bias among bureaucrats stronger than that bias toward the rectitude of positions taken in the past. A myriad of interlocking policies and positions taken in the State Department, Department of Defense, and the Arms Control and Disarmament Agency would require drastic revision if the U.S. approach to arms control is to be based on a search for Assured Survival rather than for a perpetual balance of terror.

Of immediate concern in the area of arms control are those treaties which address the uses of space and strategic defensive systems—the Outer Space Treaty, negotiated under UN aegis in 1967, and the ABM Treaty between the U.S. and USSR signed in May 1972. (Pertinent extracts from these agreements are in the annex to this chapter.)

With regard to the UN treaty on outer space, nothing in the High Frontier concept contradicts its language. The prohibition against "weapons of mass destruction" in orbit is not violated by any of the High Frontier military programs involved and the nonmilitary programs can be fairly depicted as beneficial for all countries.

Even so, the United States government would have to prepare for a polemical buffeting by the Soviet Bloc and its Third World clients for engaging in "space imperialism." The linchpin for such a propaganda assault has already been set by the Soviets in their proposed new UN treaty outlawing *all* space weaponry.

A more serious problem for High Frontier is presented by the ABM Treaty. As the only real treaty to emerge from the SALT process, it is of great symbolic value to arms control advocates. It also represents the legalistic refuge for adherents of the Mutual Assured Destruction doctrine. Finally, it was negotiated, ratified, and applauded by many influential figures from many quarters of the U.S. body politic.

High Frontier represents a direct refutation of the philosophical basis underlying the ABM Treaty. The defensive systems advocated by High Frontier do not necessarily conflict with the specific provisions of the treaty, but they can and will be construed as conflicting with both the spirit and the letter of it.

There are three basic legitimate answers to real or alleged conflict between High Frontier and the ABM Treaty: abrogate, assert compliance, or amend.

ABROGATION

The ABM Treaty provides for withdrawal by either party in the event that its "supreme interests" are jeopardized. The U.S. Senate was assured in 1972, prior to ratification, that failure to achieve progress in offensive strategic weapons limitation agreements would be grounds for U.S. withdrawal. Certainly the case can be made that SALT negotiations have failed to check the unprecedented growth of Soviet nuclear offensive power and that this jeopardizes U.S. supreme interests. Add to this the strong evidence of Soviet violations of this treaty and the case for abrogation is clear.

ASSERTION OF COMPLIANCE

The definitions of what constitutes an ABM system within the context of this treaty are rather rigid. The spaceborne ballistic missile defense systems involved in the High Frontier concept can be fairly described as "ABM systems based on other physical principles." Limitations on such systems become the subject of discussion between the signatories. Such discussion can be initiated without hindrance to U.S. action to acquire such systems.

A case can also be made, although less clearly, that certain point defense options in the High Frontier layered defense concept also fall outside treaty definitions of ABM. In any case, at least 100 U.S. ICBM silos could be protected against a first strike without violation of the treaty.

AMENDMENT

The 1972 ABM Treaty provides for review and amendment every five years. The last review in 1977 was only perfunctory. In the upcoming 1982 review the U.S. negotiating team should propose amendments to permit unfettered U.S. acquisition of defensive systems if the options of abrogation or asserted compliance are rejected or appear inadequate to support the High Frontier efforts.

CHAPTER VIII ANNEX: THE OUTER SPACE AND ABM TREATIES

LEGAL ARGUMENTS

Two international treaties currently in effect bear directly on Project High Frontier. They are the so-called Outer Space Treaty and the Anti-ballistic Missile (ABM) Treaty. The first was effected in October 1967, while the latter was signed with the Soviet Union in May 1972.

The preamble to the Outer Space Treaty refers to three United Nations General Assembly resolutions which cover . . . ''Legal Principles Governing the Activities of States in the Exploration and Use of Outer Space,'' a call upon States, '' . . . to refrain from placing in orbit around the Earth any objects carrying nuclear weapons or any other kinds of weapons of mass destruction . . . " and the condemnation of propaganda, which would " . . . provoke or encourage any threat to the peace, . . . " The Outer Space Treaty is essentially consistent with these United Nations General Assembly resolutions.

KEY POINTS OF OUTER SPACE TREATY

> The exploration and use of outer space, including the moon and other celestial bodies, shall be carried out for the benefit and in the interests of all countries . . .
>
> States Parties to the Treaty shall carry on activities in the exploration and use of outer space . . . in accordance with international law . . .
>
> States Parties to the Treaty undertake not to place in orbit around the Earth any objects carrying nuclear weapons or any other kinds of weapons of mass destruction, install such weapons on celestial bodies, or station such weapons in outer space in any other manner.
>
> The moon and other celestial bodies shall be used by all States Parties to the Treaty exclusively for peaceful purposes.
>
> States Parties to the Treaty shall bear international responsibility for national activities in outer space . . .
>
> If a State Party to the Treaty has reason to believe that an activity or experiment planned by it or its nationals in outer space . . . would cause potentially harmful interference with activities of other States Parties in the peaceful exploration and use of outer space . . . it shall undertake . . . consultations before proceeding with any such activity or experiment.

The preamble to the ABM Treaty ties defensive and offensive weapons limitations together with such language as: '' . . . the premise that nuclear war would have devastating consequences for all mankind''; '' . . . effective measures to limit antiballistic missile systems would be a substantial factor in curbing the race in strategic offensive arms and would lead to a decrease in the risk of outbreak of war involving nuclear weapons''; '' . . . measures with respect to the limitation of strategic offensive arms, would contribute to the creation of more favorable conditions for futher negotiations on limiting strategic arms.'' It declares as an intention of the ABM Treaty: ''to achieve at the earliest possible date the cessation of the nuclear arms race . . .'' and '' . . . general and complete disarmament''. . .

KEY POINTS OF ABM TREATY

> Each Party undertakes not to deploy ABM systems for a defense of the territory of its country and not to provide a base for such a defense, and not to deploy ABM systems for defense of an individual region except as provided for in Article III of this Treaty.

Describes the purpose of the treaty to counter strategic ballistic missiles in flight trajectory. Clearly identifies an ABM system, for purposes of the treaty, as (1) ABM interceptor missiles, (2) ABM launchers, and (3) ABM radars.

Limits deployment of one ABM system around a " . . . Party's national capital . . ." and another ABM system to deploy around ICBM silo launchers.

Each Party agrees " . . . not to develop, test, or deploy ABM systems or components which are sea-based, air-based, space-based, or mobile land-based." Limits the launching of one ABM interceptor missile at a time for launchers.

Each Party agrees " . . . not to deploy in the future radars for early warning of strategic ballistic missile attack except at locations along the periphery of its national territory and oriented outward."

The 1974 summit protocol further limits the number of ABM sites to one in each country. (The U.S. has none, having abandoned the Grand Forks site. The Soviets defend Moscow with their one site.)

The ABM Treaty narrowly defines the ABM system as interceptor missiles, launchers, and radars. This "tight" definition was addressed by the U.S. and Soviet heads of delegation on May 26, 1972, the same date as the original Treaty signing, in "Agreed Statements."

> In order to insure fulfillment of the obligation not to deploy ABM systems and their components except as provided in Article III of the Treaty, the Parties agree that in the event ABM systems based on other physical principles and including components capable of substituting for ABM interceptor missiles, ABM launchers, or ABM radars are created in the future, specific limitations on such systems and their components would be subject to disagreement in accordance with Article XIII and agreement in accordance with Article XIV of the Treaty.

Articles XIII and XIV refer to meetings of a Standing Commission to consider measures bearing on the Treaty as well as establishing the consideration of amendments proposed by either party. Parties will conduct a review of the Treaty every five years. (Treaty discussions would normally be undertaken in 1982.)

APPENDIX A: MILITARY AND ECONOMIC RISKS

In the early years of the nuclear era, the U.S. could rely on massive assured retaliation to deter nuclear attack by ICBMs because the credibility of such retaliation was very high. We held a decisive superiority in nuclear weapons and delivery systems. Since no means then existed to destroy ICBMs after they were launched, deterrence was a viable policy option for the U.S.

The USSR's response was to build a large ICBM force of its own. While the Soviets did this during the late 1960s and 1970s, we stood still. We hoped that through SALT, detente, and diplomacy the USSR could be persuaded to halt their developments at "parity," that a viable "balance of terror," known as mutual assured destruction (or MAD), could be thereafter maintained.

Unfortunately, the USSR did not stop at parity. They have surpassed our strategic capability by a substantial margin, and there is no evidence that they intend to abate current high production rates. With the high accuracies they have demonstrated with their ICBMs and the ever larger inventory of weapons that they are creating, their ability to undertake decisive unilateral military action against the U.S. will grow ever greater—unless we counter their initiatives. Countering this threat is undoubtedly the most important military imperative of this decade.

Fortunately, we have an option superior to trying to catch up with the Soviet proliferation of nuclear weapons. We have the technological capability to destroy Soviet ICBMs, or reentry vehicles (RVs), before they reach an intended U.S. target. For example, point defenses capable of keeping RVs out of critical range of our hardened Minuteman silos are available. The design of space based platforms with simple interceptors to kill ICBMs during boost exists. These vehicles, with improved sensors, could also be used against RVs in mid-course trajectory.

Ultimately, a three-tiered defense in which attacks on Soviet ICBMs are initiated while they are in boost phase, continues against leakage that gets through the boost phase defenses, and finally culminates with point defense (around ICBM silos and other key hardened facilities) can be ours of we choose to create it. Such a defense would pose a formidable barrier to any Soviet consideration of a surprise nuclear strike.

The following pages outline some of the possible consequences of the U.S. pursuing "Assured Survival" through missile defense. Comparisons will also be provided with building new systems such as MX.

The study concludes that reduction of the military risks we face is mandatory—hence funding of ballistic missile defenses is a true *imperative*. Such funding is clearly of higher priority than funding *any* other kind of incremental increase in military capability.

U.S. RESPONSE TO USSR STRIKE

Our principal alternatives in reacting to a Soviet first strike range from choosing to accept passively such a strike (absorb the consequences), to launching on warning, to actively defending against the attack.

DEFEND

In this study, defense will be delineated as point defense, boost phase defense, or mid-course defense. For example:

POINT DEFENSE

Point defense means close-in defense of Minuteman silos only. Radar would be employed to trigger the RV kill mechanisms (intelligently guided rocket powered interceptors, "swarms" of small projectiles or flechettes or possibly high fire-rate guns). Since they are nonnuclear, these could be autonomous systems (would fire automatically at any very high speed object coming in on a predetermined course). Kills could occur at ranges as close as three to four thousand feet—hence decoys would not be a problem. (They would have burned in reentry.)

BOOST PHASE DEFENSE

Space based platforms each with 30 to 150 guided interceptors would be continuously circling the Earth in orbits that would place them over the desired Soviet target areas for only a fraction of each orbit—hence many platforms would be required. Between 200 and 500 are contemplated. Consequently, several thousand interceptors would be necessary in order to have about 1000 always usefully close to the Soviet launch complexes. Boost phase interceptors would use readily available infrared sensors to home in on the heat of the ICBM's propulsion units.

MID-COURSE DEFENSE

Since the space based system above would have hundreds of extra interceptors which would come into useful ranges while the RVs are in mid-course transit, these interceptors would be available to kill RVs that elude the boost phase interdiction effort. Mid-course intercept, however, may require the development of improved sensors to see the small and now cooler reentry vehicles and provide intercept guidance.

ABSORB USSR FIRST STRIKE

If an international situation should develop in which the gains to be gotten would, in the Soviet view, justify a surprise attack, what would be the consequences for the U.S., if we choose not to have missile defenses in the future?

Assuming an 80 percent net kill probability against Minuteman, the Soviets need target only two RVs per Minuteman silo to achieve a 96 percent annihilation of our Minuteman force. In numbers, only about 40 of our 1000 + Minuteman force would survive.

A parallel attack on 1000 other softer targets of military value would result in only one percent or 10 of these surviving.

LAUNCH ON WARNING

If the Soviets view this option as a credible policy, it could be a strong psychological deterrent. It could buy us valuable time while we build a ballistic missile defense. However, in the event of a Soviet first strike, nonetheless, it is a hollow policy. If we succeed in launching on warning (we succeed in obtaining launch authority from all levels concerned in a very short time—even if Soviet ICBM launch is not positively verified) then there will be a massive exchange. Otherwise, we absorb a unilateral first strike as previously discussed.

In either event, the effectiveness of almost all of our land based military resources would be nullified by the incoming Soviet attack.

THE EFFECTS OF GROUND BASED POINT DEFENSE ONLY

Using existing technology and off-the-shelf components, point defense of 20 percent of Minuteman could be rapidly implemented. If such systems employed simple swarms of small rockets or flechettes, the kill probability of these systems might be as low as 0.4. Nonetheless, with this limited deployment and low effectiveness, the survivability of Minuteman would be *more* than double the previous case (86 versus only 40 surviving previously). If more effective point defense systems are implemented (multiple shot systems with high kill probabilities) then Minuteman survivability exceeding 70 percent can be postulated against Soviet attack (see Figure 15).

MINUTEMAN SURVIVING

		Percent MM Protected	
		20%	100%
Defense Kill Probability	.40	86	270
	.80	174	706
Other Targets	10		

Figure 15. Ground-Based Point Defense

THE EFFECTS OF LAYERED SPACEBORNE DEFENSE

Figure 16 illustrates the value of a possible space based missile defense system. Even if such systems exhibited kill probabilities of only 50 percent, these systems would assure that over one-third of the Minuteman force would survive. In actuality, preliminary engineering studies suggest that kill probabilities exceeding 90 percent are possible. Consequently, Minuteman survivability, when protected by a defense with a somewhat more conservative 0.85 kill effectiveness is also shown.

The survivability of soft targets may be fully as important as having our missile retaliatory forces survive because it is most unlikely that a Soviet ICBM attack would not be coordinated with ground forces moving toward preselected objectives.

	Surviving Minuteman	Other Military Targets
$P_{KILL} = 0.5$	360	302
$P_{KILL} = 0.85$	774	748

Figure 16. Boost Phase Interdiction

In Figures 17 and 18 following, the effectiveness of defense in depth is estimated. With mid-course interceptors taking a toll of any ICBM or RVs escaping the first defensive line, Minuteman survivability jumps to over 90 percent with effective system design.

In the charts following, a triple layered defense including boost, mid-course, and terminal or point defense (for Minuteman only) is depicted. The cumulative effectiveness of such defense should totally deter nuclear attack via ICBM, either as a surprise or as the consequence of escalation to theater level or higher warfare.

	Surviving Minuteman	Other Military Targets
$P_K = 0.5$*	640	600
$P_K = 0.85$*	964	960

*Both boost phase and mid-course interdiction

Figure 17. Boost Phase Plus Mid-Course Interdiction

	Surviving Minuteman	Other Military Targets
Low P_K	774	600
High P_K	993	960

Figure 18. Point Defense Plus Space Defense

SURVIVABILITY—ALL MILITARY TARGETS (see Figure 19)

Space based defense provides survivability to *both* our retaliatory missile forces and our conventional military forces. The military value of the survivability of these forces is incalculably high.

First, the likelihood that we would survive a Soviet surprise nuclear attack with the majority of *both* our missile and conventional forces intact should preclude any Soviet attempt to perpetrate such an attack. Therefore, these defenses constitute a *positive* deterrence to possible Soviet opportunism—a deterrence that would not be based on Western intellectual presumptions concerning the unacceptability of nuclear warfare to the Soviet Politburo or military leaders.

Second, in the event that limited or wide scale nuclear war does ensue, we will have maximum chances of using our conventional military forces during and after any exchanges. Since nuclear warfare is the worst possible deterioration of military confrontation, it would only be undertaken if an extremely valuable objective were to be gained. Therefore, it is presumed that the exchange must be followed by conventional force moves to capture the desired objectives. Consequently, the retention of maximum general purpose military capabilities *after* nuclear attack would appear to be an imperative of highest order.

Finally, the existence of defenses creates uncertainty—uncertainty that a Soviet strategic planner cannot resolve. The military value of such uncertainty could be pivotal because this factor could preclude the Soviets even considering many options that they might otherwise find to their advantage.

DOD EXPENDITURES AT RISK (See Figure 20)

To put survivability in an economic context, consider how much the U.S. has spent on defense over the past decade. To equip and maintain an Army with three-quarters of a million trained men has cost the U.S. a quarter of a trillion dollars (current dollars) over the last decade. Our

	Units Existing (1981)	Surviving Units (High P_K's)				
		MX-MPS (200 MX) (No Defenses)	100% Point Defense	Boost Phase Defense	Boost & Mid-Course Defense	3-Tier Defense
	Ref.	1	2	3	4	5
ICBMs	1000	45	706	774	964	993
Air Bases	140	1	1	105	134	134
Army Bases	60	1	1	45	58	58
Naval Bases	70	1	1	52	67	67
Overseas Bases	140	1	1	105	134	134
Other (Nuc., Fuel. . .)	590	6	6	441	567	567
Total	2000	55	716	1522	1924	1953

Figure 19. Survivability—All Military Targets

Navy with its ships and all its associated men and equipment has cost a little more—about one-third of a trillion dollars. These services, plus the Air Force at intermediate cost, have cost us nearly $900 billion in toto through the 1970s. This is the expense we have borne in order to be able to defend ourselves should the need arise. However, if the Soviets elect to commit to a nuclear strike, this past investment on our part would be largely nullified. With the ability to destroy 99 percent of all bases they choose to hit, they could summarily destroy the majority of our facilities and their corps of trained men.

In simple terms, if we choose not to buy ballistic missile defense (BMD) at $30 to $50 billion of cost, the hundreds of billions that we have spent to have this broad spectrum of defense forces can be destroyed at will by Soviet military action if the stakes of the game should justify that level of military risk.

CURRENT DOLLARS

ARMY		NAVY	
16	Divisions	531	Ships
758,000	Trained Men	686,000	Trained Men
60	U.S. Bases	70	Bases
$255B	Cost (1970s)	$323B	Cost (1970s)
$645B	Cost (1980s)	$817B	Cost (1980s)

AIR FORCE		TOTAL DOD	
130	Squadrons		
559,000	Trained Men	$870B	Services
140	U.S. Bases	$124B	Other
$292B	Cost (1970s)	$994B	Cost (1970s)
$739B	Cost (1980s)	$2515B	Cost (1980s)

Figure 20. DOD Expenditures

As the current defense budget takes hold, the amount of DOD spending at risk will rise to greater heights. For example, about $2.5 trillion dollars in total DOD expenditures is contemplated through the 1980s. Is it conceivable that the U.S. would forego an expenditure of only $30 to $50 billion (from this $2500 billion amount) to protect all the defenses to be purchased at this cost?

U.S. LOSSES AT REPLACEMENT COST

Figure 21 compares our options if one considers only the replacement cost of the facilities and equipment at risk in the event of a Soviet strike.

The first column is a rough estimate of the value of all the targeted facilities plus equipment such as planes or ships usually likely to be there priced at replacement cost in 1981. These military facilities would cost about $725 billion if totally replaced. If the U.S. builds MX-MPS, the total investment lost (column two) to a successful Soviet strike (if we do not have ballistic missile defenses) comes to $764 billion—more than the a priori value of $725 billion because the value of MX investment destroyed in the strike is also included.

For all the defense cases following only high kill capability systems are considered.

Point defense alone (100 percent coverage of Minuteman with a P_K of 0.8) reduces the amount of Minuteman lost to about one-third of its total cost, but point defense suitable for hard sites only does not save any of the huge losses postulated for the soft targets.

With space based defense the large DOD investments in bases and other targets can be protected—particularly if both boost and mid-course interdiction are implemented. Facility and equipment losses decline to under five percent in this case.

The final case, three tiers of defense for Minuteman, cuts Minuteman dollar losses to under a half a billion dollars. However, it would not seem to be cost effective to spend $10 billion to trim only about $2 billion of Minuteman loss. This would be true if point defense were added *after* the space defenses were in place. In actuality, point defense is essential as the first system implemented, for it can be implemented far more rapidly than the more effective space systems. This is essential if we are to reduce our near term vulnerability which is particularly acute, as previously discussed.

CONCLUSIONS

The Soviet Union has a nuclear strategic capability and an ongoing production capacity (with no cutoff in sight) which permits them in this decade to threaten—or to use at will—to destroy the majority of our land based retaliatory and conventional military forces. Our vulnerability will be particularly acute by the mid 1980s.

Virtually all of the existing defenses purchased by the DOD through past expenditures and virtually all the defense to be acquired by the enhanced programs of the current Administration will depend totally on "deterrence" working—if we choose not to implement ballistic missile defenses.

Defense sufficient to enhance Minuteman survivability by several factors can be purchased for far less than buying equal capability through expansion of our missile forces. Point defense of Minuteman can also be implemented quickly if procured by expedited procedures if effectiveness requirements are set by current technology.

Effective defense for all military targets (as well as U.S. cities and civil targets) can be acquired through space based systems of moderate cost. These systems are the only means known, short of keeping pace with Soviet missile proliferation, for denying the Soviets unilateral first strike capability.

The seriousness of our vulnerability dictates that the U.S. act. Implementation of defenses immediately based on available technology is critical to our survival. Virtually all DOD expenditures planned are subject to the given threat, consequently, none can be considered more important than expenditure to counter that threat.

MILITARY TARGETS

	Facilities Value (Replacement Cost) 1982$	Cost of Facilities Lost				
		MX-MPS (200 MX)	100% Point Defense	Boost Phase Defense	Boost & Mid-Course Defense	3-Tier Defense
Case	Ref	1	2	3	4	5
Minuteman	55	102	16	12	2	0
Air Bases	240	238	238	60	10	10
Army Bases	70	69	69	18	3	3
Naval Bases	150	148	148	38	7	7
Overseas Bases	40	40	40	10	2	2
Other (Nuc., Fuel, Etc.)	170	167	167	43	7	7
Totals	725	764	678	181	31	29
Defense Cost		50	10	25	30	40

Figure 21. U.S. Losses at Replacement Cost—Military Targets

APPENDIX B: TYPE POINT DEFENSE SYSTEM

The basic system involves three major components: (1) a radar system located 10,000 to 20,000 feet forward of the Minuteman silo to detect, track, and calculate the optimum intercept point for the incoming reentry vehicle; (2) a launcher system which is hardened to protect itself against a one-megaton blast at a range of 3,000 feet and which aims and launches rockets in times on the order of one second; and (3) a swarm of ballistic rockets (about 10,000 in number) which fly to the intercept point at velocities on the order of 5,000 feet per second and kill the reentry vehicle with the kinetic energy of impact from a single rocket case striking it (see Figures 22 and 23).

RADAR SYSTEM

The study showed that an acceptable radar system would include an array of three low cost radar stations (UHF, VHF, or X-band) which determine the reentry vehicle track and intercept point by a trilateration scheme. For the trilateration deployment, the radars can be low cost because only range information is required. To prevent the necessity of hardening the radar against direct attack, a trilateration array should consist of four radar stations, and each Minuteman silo should be protected by two arrays. The radars in each array should be located far enough apart so that only one radar can be knocked out by a single nuclear warhead. Thus, it takes a minimum of two (potentially four) reentry vehicles to knock out the radar.

The radar need not search above 40,000 feet altitude in order to provide sufficient tracking

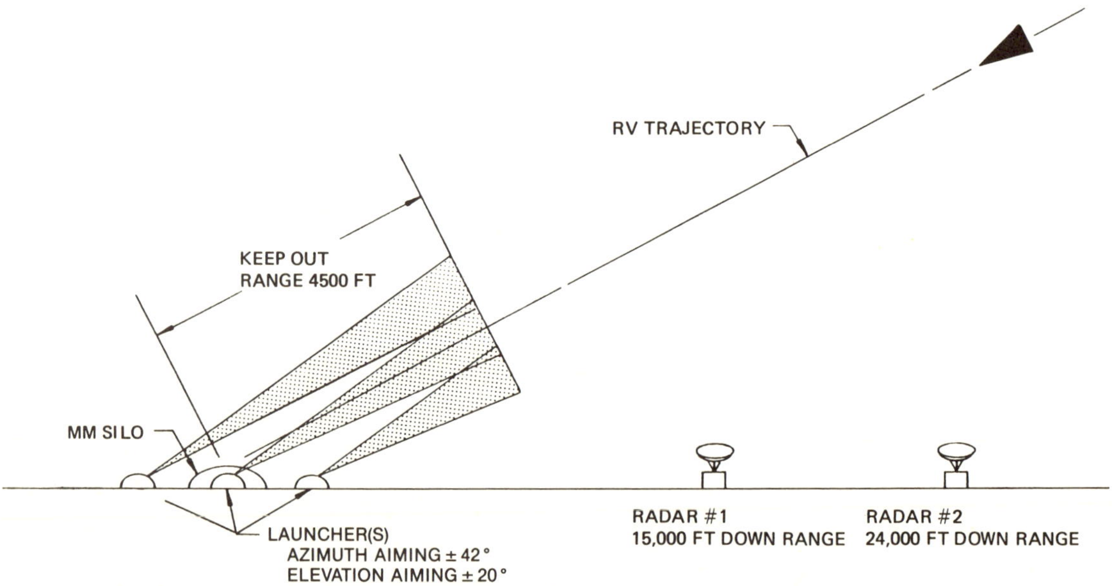

Figure 22. Swarmjet Engagement Schematic

time (two seconds) to predict accurately the intercept point. Thus, the radars are not affected by high altitude blackout, decoys, or clutter which may accompany the reentry at high altitudes. Because two arrays are located forward of the silo, there is no problem in looking around a nuclear fireball or even a low level nuclear blast and detecting other incoming reentry vehicles.

LAUNCHER SYSTEM

The launchers could be hardened by either a concrete bunker with removable doors or individual steel shells, since the static overpressure from a nuclear blast at 3,000 feet is only 150 pounds per square inch. The launcher slew rate required (1.5 radians per second) is well within the conventional launcher capability, and the aiming precision required (two to three mils) compares favorably with the one-mil precision often built into gun systems. Since each launcher would contain 500 to 1,000 rockets, the preferred design is an open tube (recoilless) launcher to minimize the stress on the structure and on the launch tube weight.

SWARMJET PROJECTILES

The unguided rockets which make up the swarm are spin stabilized like a bullet. Their diameter should be between one and three inches and their length 10 to 15 inches. Using cases of conventional steels and one of the better ammonium perchlorate propellants, the kinetic energy of the impact of the rocket case with the incoming reentry vehicle could exceed that required to penetrate and destroy the warhead at ranges on the order of 4,000 feet.

The number of rockets required to give an 85 percent probability of intercepting the reentry vehicle in its lethal area, considering the uncertainty in ascertaining the position of the reentry vehicle and the uncertainties in the aim and flight of the rockets, was between 5,000 and 10,000 rockets if the intercept point was 4,500 feet from the silo.

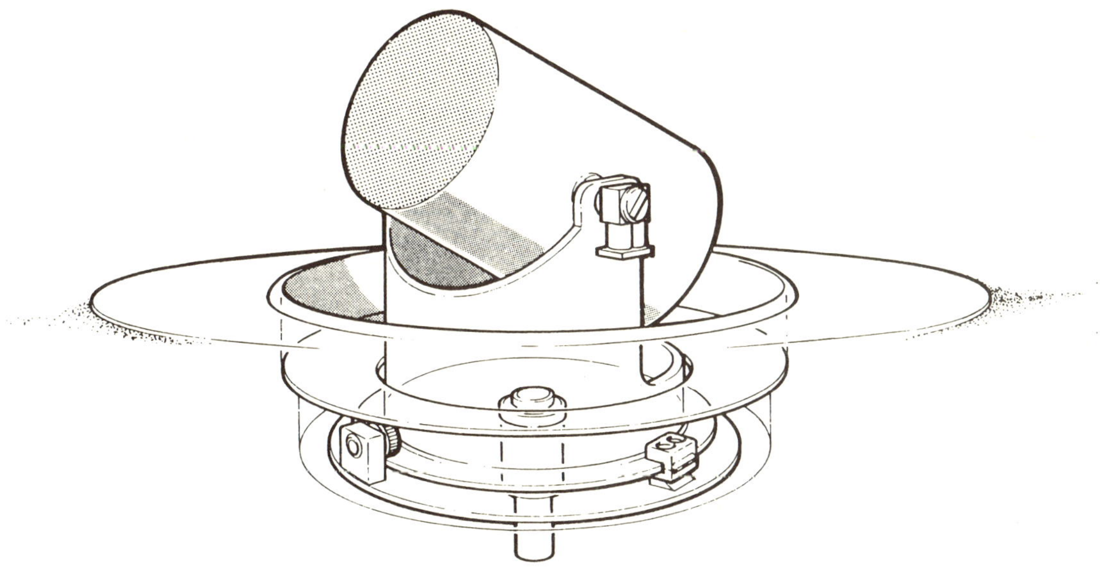

Figure 23. Hardened Swarmjet Launcher

SYSTEM COSTS

The manufacturing phase of the study showed that the conventional propellant casting process could be accelerated and that by gathering available facilities into an ordnance type production line, production rates of five to ten million rockets per year could be achieved. This would be adequate to provide the rockets to defend 500 silos with three swarms of 10,000 rockets each (15 million total), within the required timeframe. Based on using available launch drive systems, concrete bunkers for hardening, and 1,000 rockets per launcher, system costs are approximately $4 billion, including five years operation and maintenance, or $8 million per silo. This would defend each silo against the first three arriving warheads. For defense against two incoming warheads, the cost per silo is about $5 million.

STRUCTURED ATTACK

One area of system concern is that of handling a highly structured, multiple reentry vehicle attack in which the enemy can space the incoming reentry vehicles five to ten seconds apart and equip the reentry vehicles with salvage fusing or deliberately predetonate the first in the series. The winds generated by the first nuclear warhead generate a window of five to ten seconds, during which time it is unlikely that a successful SWARMJET launch could be made. A reentry vehicle entering through the fireball of the first reentry vehicle during this period could probably successfully attack the silo. Such a structured attack would require detailed knowledge of the system operation, a commitment by the attacker of four or more weapons to each target attacked, and the perfection of highly sensitive salvage fusing of reentry vehicles.

APPENDIX C: GLOBAL BALLISTIC MISSILE DEFENSE

GENERAL

This spaceborne ballistic missile defense concept is designed for the boost and post boost nonnuclear interception of ground based, sea and air launched ballistic missiles such as the ICBM, IRBM, MRBM, and SLBM. It can also be used for boost and post boost interception of other rocket boosted vehicles such as the communication satellite, intelligence satellite, and the assessment satellite. The global ballistic missile defense (GBMD) system includes a large network of satellites that are launched in peacetime by launch vehicles that comprise the first three stages of the MX ICBM booster. The network may carry tenant systems for uses other than ballistic missile defense. Potential tenant systems include command, control, communication, and intelligence (C^3I), nuclear detection, tactical warning, and attack assessment systems. As an area defense system it defends forces other than ballistic missiles inherently. Capable of defending cities and countries, the GBMD system may provide defense with international financial and political support.

Heretofore, ballistic missile defense (BMD) programs and concepts have been ground based and thus limited by the following characteristics and requirements:

- Defense of missile sites.
- Interception of each reentry vehicle (RV).
- Interception of RVs during the final phase of their trajectories.
- Need to discriminate between RVs, decoys, and other objects in free-fall.
- Operation in severe nuclear environment.
- Dependence on optical and radar transmissions through the atmosphere in a hostile environment.
- Lack of realistic test capability.
- Inability to defend other forces, cities, or countries.
- Inability to host other systems.
- Costs must be added to the ICBM and basing costs.

These limitations and characteristics contrast sharply with the flexibility, performance, and cost effectiveness potentially achievable by the space based GBMD system presented herein.

Any spaceborne defense system must be on-orbit survivable. Launched in peacetime, the ground launch facilities do not have to be hardened against attack. A single launch point may be sufficient for survivability as contrasted with the multiplicity of launch points for the ICBM or the ground launched BMD interceptor. The on-orbit elements of the system must be survivable in peacetime, conflict, and war up through levels of nuclear war and be capable of enduring all these conditions with a demonstrable performance capability residual. This is not to state that all elements of the system or even perhaps the great majority of the elements must survive. The key point is that the system must be sufficiently survivable to provide an effective filter of a ballistic missile attack.

The GBMD system should be deployable in the near term, within a period of approximately five years. It should have a reasonable life cycle cost and life cycle effectiveness. Life cycle effectiveness correlates directly with survivability.

The GBMD system is required to intercept its target during the target's boost or post boost phase. The point is to negate as many RVs as possible. Where RVs are MIRVed, post boost intercept will prevent subsequent deployment of RVs.

Kill is by nonnuclear impact at very high relative velocities similar to the intercepts planned with the Air Force's antisatellite miniature vehicle (MV) program and the Army's homing interceptor (HIT) vehicle.

The system must not be dependent on the survival of ground based C³I or warning. This is not to say that such ground based systems will not be used by the GBMD system. Rather, it is to assure the capability of operation in autonomous modes. This requirement recognizes the large uncertainties involved in our knowledge of C³I during nuclear war.

To exploit the unique opportunity of the orbital system provided by its location, line-of-sight distances and remoteness, other uses such as C³I, tactical warning and attack assessment, verification, and nuclear event detection may be provided by the GBMD system. Such systems are termed tenant systems in this report. The requirement, therefore, is to provide whatever is practicable for the implementation of tenant systems to enhance land, sea, air, and space forces.

The Soviet Union has deployed its ICBMs, IRBMs, and MRBMs across its country as is indicated in Figure 24. The missiles are also distributed rather uniformly from north to south in western Russia, while in eastern Russia the missiles are deployed relative to the Trans-Siberian Railway. The GBMD system will be shown to be insensitive to the location, number, and distribution of the target missiles before launch. Mobile missiles and tactical or strategic reserve missiles will not evade the GBMD system simply by their location uncertainty, number, or distribution.

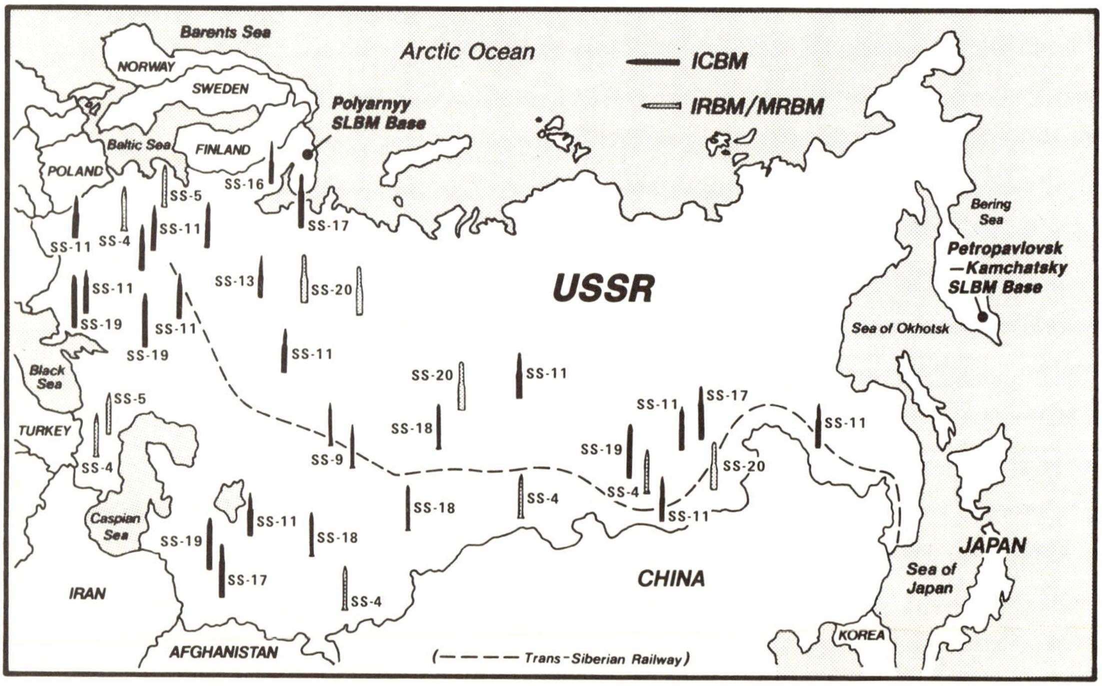

Figure 24. Soviet Ballistic Missile Distribution

In this report the SS-18 ICBM is referenced against the GBMD system. These are shown located between Tyuratam and Novosibirsk. The SS-18 represents the greatest threat because it is accurate, can deploy at least 10 multiple, independently targeted RVs, and is being deployed in large numbers.

SYSTEM DESCRIPTION

A representative system includes a large network of these satellites or "trucks" distributed in circular orbits at an altitude of approximately 300 nautical miles (nmi). The example referenced herein uses 432 trucks, all in orbits inclined 65 degrees with the equator. Missile booster and bus signatures are observed from the trucks at optical wavelengths appropriate to tracking against the Earth as a background. The truck contains and can deploy 40 to 45 self-propelled objects called carrier vehicles (CVs) which are each capable of obtaining a velocity with respect to the truck of 3,000 feet per second. The velocity is a critical factor, on which depends the kill capabilities designed into any particular GBMD concept. Furthermore, the costs of the system are very sensitive to the velocity capability of the kill vehicle. The truck can be capable of tracking and supervising the control of each CV during its trajectory to intercept.

Each CV includes a propulsion module (PM) and the kill vehicle (KV), which can be separated from the PM prior to intercept and after the KV has established optical tracking of its target. Each CV will have a pulsed light source whose pulse train identifies the CV uniquely. Thus, a number of CVs can be in the truck's field of view without causing ambiguity. This is the ''traffic problem'' solution. The truck will be capable of commanding each CV with signals for mid-course corrections, target designation, intercept inhibition, deorbit and burnup, etc.

Clearly, multiple strikes or intercepts can be made from a single truck against one or a number of targets. Multiple deployment will increase the kill probability.

INTERCEPT GEOMETRY

If the CVs were deployed simultaneously in all directions their distribution would be characterized as an expanding, approximately spherical surface moving with and centered on the truck which is moving in free-fall along its orbit.

Let us consider a representative cross-track intercept and make additional observations. Figure 25 depicts the intercept of an SS-18 missile at the end of its boost from Tyuratam by a truck located over Saudi Arabia. Interception is indicated at about 350 seconds from truck deployment, corresponding to carrier vehicle deployment about 53 seconds prior to actual missile launch, when the truck is about 950 nmi ground range from the missile launch point. If the truck were to move along its trajectory for 50 seconds it could deploy carrier vehicles for final stage intercept in response to direct viewing of the missile launch. Ground range separation between the missile and the truck would be about 660 nmi for this case and the truck would be located over Iran. Trucks in orbits east of the orbit shown will similarly be able to intercept targets with CV deployments while over Afghanistan and China. Trucks in orbits to the west of the orbit shown may deploy carrier vehicles for intercept while the trucks are over the Mediterranean Sea, Turkey, or European countries. These interceptions would occur during the post boost phase and would permit some RVs, deployed before intercept, to leak through the GBMD system. Study of Figure 26 shows that trucks may also intercept the Tyuratam missile near the end boost point, in essentially a head-on approach. These trucks are traveling down and to the right toward decreasing latitudes.

When a truck has moved along its orbit for 200 seconds (in time), it will have advanced approx-

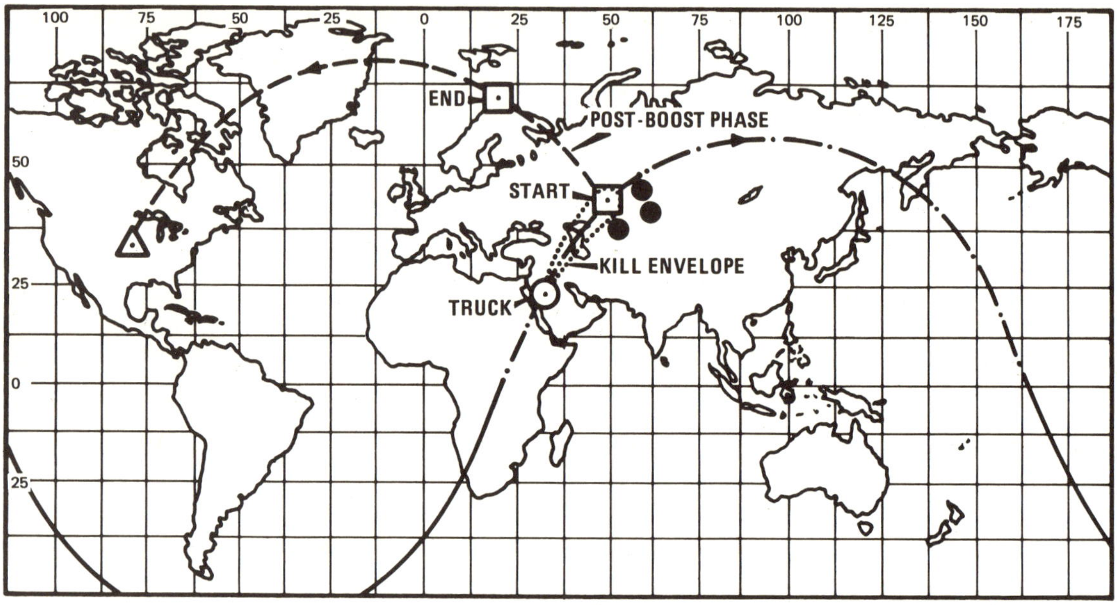

Figure 25. Cross-Track Intercept Example

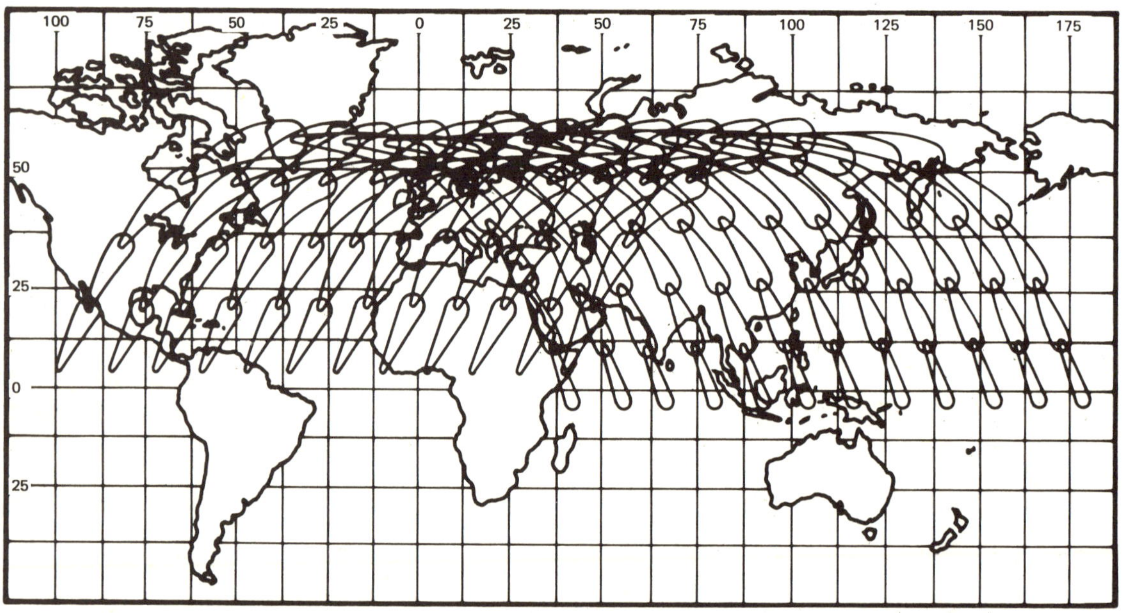

Figure 26. Example Footprints (1/2 Northern Hemisphere)

imately 12.5 degrees along its orbit. The trace of the expanding envelope of the deployed carrier vehicles is indicated by the dashed-line footprint indicated in Figure 25. The footprint shown represents footprints at 50 nmi above or below the truck at 300 nmi.

SYSTEM DEPLOYMENT

Figure 26 introduces the consideration of the complete global pattern of footprints. For clarity, only half of the footprints in the Northern Hemisphere are shown and all footprints over the Southern Hemisphere have been omitted. Each represents the plan view of the footprint at an altitude of 250 nmi or 350 nmi. Of course, the footprint at 300 nmi extends to the truck. The apparent differences in footprint area are the consequence of the particular cartographic projection (Miller cylindrical) and do not imply a real difference in footprint area among the trucks.

The 65-degree orbit was selected for illustrative purposes because it allows the presentation of several points during the discussion that would be missing if a polar orbit were used. Twenty-four orbits equally spaced at 15-degree longitudinal increments along the equator constitute the full set. In this particular example, which does not exclude other possible orbital geometries, each orbit contains 18 trucks spaced in 20-degree increments. Thus, a total of 432 trucks comprises the full GBMD exemplar. Note that the 65-degree orbit causes substantial footprint overlap in the 45 to 65-degree latitudes. Small gaps or holes are seen among the footprints which increase in area with decreasing latitude to the equator. These holes pulsate in size, as a function of time. The central pattern of the figure would extend uniformly across the hemisphere if all system footprints were shown. The complete pattern would be duplicated across the Southern Hemisphere as well.

Reference to Figures 25 and 26 illustrates that the GBMD system has a second opportunity to intercept a Soviet strike against the United States by interception of RVs prior to their entering the atmosphere. The RVs will be at risk while traversing truck footprints over North America, approximately 30 minutes after launch. The footprints over Canada and the northern United States continue the overlapping characteristics of those over Europe and the Soviet Union. The trucks in position to intercept these RVs will be located over the Pacific when the missiles are launched in the Soviet Union. It is recognized that it is inherently more difficult to intercept the free-falling RVs without the plumes and heating effects of propulsion associated with booster and bus intercepts. It is also fair to point out that the vulnerability of the trucks attempting to intercept the RVs may be less than that of those operating over the Eastern Hemisphere.

LAUNCH

The truck will be launched by the three stage booster of the MX ICBM. The truck replaces the MX bus and acts as the necessary fourth stage to propel the truck to orbital altitude and to provide the insertion velocity to circularize orbit. Subsequently, the truck's propulsion and reaction control system will provide stationkeeping or orbit-adjust maneuvers and evasive maneuvers if required to avoid interception. The truck contains a storable propellant and restartable liquid propulsion system with an axial propulsion engine.

It is expected that the trucks would be launched from a single facility on an island in the Pacific relatively close to the equator, such as Kwajalein, to take advantage of the Earth's rotation tangential velocity and to increase the payload to orbit over that obtainable with launch into the same orbit from the higher latitudes of the continental United States. Rapid-fire sequential launches should not be difficult. Indeed, submarine launched ballistic missiles can be launched at firing rates which are greatly in excess of GBMD requirements. The GBMD launch site need not be hardened because the trucks are launched in peacetime.

TENANT SYSTEMS

Once this system of interceptor satellites is in orbit—or, for that matter, once any sizeable portion of them is in orbit—a sort of "geodesic dome" of satellites is formed, essentially encompassing the globe. These satellites will be constantly gathering information with their sensors and passing that information among them and to ground control station. This creates an opportunity to make the GBMD system doubly valuable as a C^3I system. If the GBMD system is as survivable as the High Frontier study concludes it to be, this potential added capability can solve another of the gravest U.S. security concerns, the vulnerability of C^3I.

SURVIVABILITY

The on-orbit elements of the system must be survivable through levels of nuclear war and be capable of enduring with a sufficient performance capability residual. This is not to state that all elements of the system or even perhaps the majority of the elements must survive. The key point is that the system must be sufficiently survivable to provide an effective filter of ballistic missile attacks. The number of interceptors required for GBMD to be effective will be a function of such things as whether there is a second tier, terminal defense system (for example) and the number of U.S. ICBMs which must survive the attack.

There are passive as well as active defense options that must be considered. Figure 27 presents in matrix format a row list of potential GBMD defense options or counter-countermeasures (CCM) versus a column list of possible Soviet countermeasures (CM). To be realistic, other considerations must be included in Figure 27 determining whether a countermeasure presents a significant threat to the GBMD. Among these are:

- Level of Soviet technology required.
- Time to initial operating capability.
- Cost and other resource availability.
- Military impact and acceptance.
- Political impact and acceptance.

COST PARAMETERS

While eventual overall costs have not been estimated, it appears likely that those costs will need not be excessive. This is because a GBMD can use off-the-shelf components or technology and does not require highly accurate guidance or hardened launch facilities.

The use of already tested and developed components takes maximum advantage of sunk-cost of previous military and space research and development. This permits the rather straightforward estimate of costs for hardware using unit prices provided by probable suppliers. The results of that approach for an illustrative GBMD system are portrayed below.

In addition to the above costs, a dedicated launch facility cost is estimated to be a $500 million investment. Two ground C^3 stations, at $100 million each, add $200 million to the cost, giving an investment total of $13.3 billion for the system.

Operating costs should run about $60 million per year, without replacement or maintenance in space. An operating life of 10 years appears to be reasonable. Consequently, the actual operating costs per year must be increased by a judgment on the space maintenance and repair costs and/or possible replacement of trucks and their loads.

This $13 billion for 450 on-orbit space vehicles appears startlingly low when compared with unit costs for current U.S. space systems. This is because the unit costs of the current types of U.S. reconnaissance or communication satellites include research and development costs. Further, to

CCM / Countermeasure	I Shoot-Back	II Netting	III Survival Aids	IV Hardening*	V Maneuver	VI Proliferation*	VII Replenishment	VIII Other Comments
A. Air-Launched ASAT	1. SOP 2. GBMD insensitive	1. SOP	1. Decoys 2. ECM	1. Potential – Weight limited to low (kinetic) energy attacks	1. SOP – Translations: Cross-range or along orbit (time)	1. Feasible – Reduce system sensitivity to individual loss	1. Feasible – Use cached CV s	1. SU can deploy A/C away from SU/PACT 2. Not evidenced 3. Tracking quality req'd 4. Small TW 5. NUC CM fratricide
B. Ground-Launched ASAT	1. SOP 2. GBMD insensitive to ASAT location	1. SOP	1. Decoys 2. ECM	1. Potential – Weight limited to low (kinetic) energy attacks	1. SOP – Translations: Cross-range or along orbit (time)	1. Feasible – Reduce system sensitivity to individual loss	1. Feasible – Use cached CV s	1. Heartland launch 2. Tracking quality req'd 3. Large TW potential 4. Limited LF s 5. Drawdown ICBM LV s & LF s 6. NUC CM fratricide
C. Ground-Based Laser	1. Not possible	1. SOP 2. Deploy CV s under attack	1. Decoys	1. Laser harden	1. Roll – versus CW laser	1. Doubtful value	1. Doubtful value	1. Pulse threat larger 2. Laser soft target 3. Locates laser (net) 4. Can proliferate
D. Space-Based Laser	1. Not possible	1. Possible 2. Deploy CV s under attack	1. Decoys	1. Laser harden	1. Roll – versus CW laser	1. Doubtful value	1. Doubtful value	1. Pulse threat larger 2. Laser very soft & few 3. Laser SOA?
E. Rifle-Shot	1. Not possible	1. Possible 2. Deploy CV s under attack	1. Decoys	1. Doubtful value	1. SOP	1. Doubtful value	1. Doubtful value	1. Truck moving 24880 FPS 100/Sec ≈Δ S = 250 Ft 2. Rifle SOA?
F. Pumped Belts	1. Not possible	1. Possible – Cooperative targeting of CV s	1. Not appropriate	1. Potential	1. Change altitude	1. Altitude, etc. differences		
G. Spoofing	1. Not possible	1. Possible – cooperative source verification & validation	1. Not appropriate	1. Not appropriate	1. Not appropriate	1. Possible – if depleted use cached CV s	1. Possible – if depleted resupply CV s only	
H. Jamming (Ground)	1. Not appropriate	1. Possible – cooperative	1. Not appropriate	1. Not appropriate *Includes super hardened box of CV s per proliferation	1. Not appropriate	1. Not appropriate *Includes replenishment by cached CV s on-orbit w/o truck complexity	1. Not appropriate	Effective against RF sensors Low power laser could be used to jam EO/IR sensors wide beam to attack many simultaneously

Figure 27. Example Matrix of Counter-Countermeasures Versus Countermeasures

ensure against failures, the normal practice is to acquire three complete satellites with three reserve launching systems for each satellite required on orbit. These sharply cost-inflationary factors do not apply to GBMD, which requires minimal R&D and profits from economy of scale. Launch or on-orbit failure of a single vehicle is not crucial to the mission of the overall system. In Chapter V it was emphasized that further design considerations may call for larger trucks with higher velocity kill vehicles. This would involve increased costs.

COST ESTIMATE—GLOBAL BALLISTIC MISSILE DEFENSE

LAUNCH VEHICLE	$8 M[(a)] / MX BOOSTER	X	1 MX BOOSTER / TRUCK	X	450 TRUCKS	= $3.6 B
CARRIER VEHICLE	$88 K / KV + PM	X	45 KV / TRUCK	X	450 TRUCKS	= $1.8 B
TRUCK	$10.4 M[(b)] / PLATFORM (TRUCK)	X		X	450 TRUCKS	= $4.7 B
SENSOR & C^3 PACKAGES	$3.3 M[(c)] / PLATFORM (TRUCK)	X		X	450 TRUCKS	= $1.5 B
RESEARCH & DEVELOPMENT						$1.0 B
TOTAL FOR SYSTEM IN SPACE						$12.6 B

[(a)] Initial runs for MX will cost $10 million each. For substantial quantities, the average cost should run about $8 million per MX. This includes the guidance package for placement into orbit.

[(b)] It is U.S. industrial experience that the minimum cost of any space platform in small quantities is about $10 million.

[(c)] This includes the sensor and C^3 packages for the truck, the CV, and KV.

APPENDIX D:
HIGH PERFORMANCE SPACEPLANE CONCEPT

A truly military, piloted, high performance spaceplane (HPSP) concept is presented for the near term accomplishment of a comprehensive set of space missions at all altitudes from within the Earth's upper atmosphere to beyond the Moon. Mission categories are exemplified by reconnaissance and surveillance; inspection and verification; antisatellite and anti-antisatellite; placement, supplementing, and standing in for unmanned satellites; on-orbit service, repair, and update of satellites; and missions requiring multiple atmospheric entry and exit. The HPSP transforms the Space Shuttle Orbiter into an aircraft carrier in space and extends its military operations throughout cislunar space (space volume between the Earth and the Moon). Compatible with ground and air launch, the HPSP can also operate completely independent of ground operations and the Shuttle system. The HPSP can be cached on-orbit, brought back in the Orbiter, or piloted to a landing at unprepared sites, airfields, aircraft carriers, or other ships. The HPSP differs considerably from the other manned and unmanned space vehicles that have been proposed or studied. It differs in configuration, cost, performance, ease, and speed of development and in launch and recovery flexibility.

THE PROBLEM

The problem is the nonmilitary characteristic and severely limited military capability of current and proposed spacecraft at a time when the military need is substantial and increasing rapidly. Manned spacecraft programs and concepts have been and are continuing to be characterized by:

- Dependence throughout their mission on extensive ground support monitoring, tracking, control, and communications.
- Extreme cost of acquiring, operating, and maintaining the ground support and launch facilities and personnel.
- Vulnerability of the launch facilities and the global ground support to direct attack.
- Severely limited space maneuverability.
- Substantially constrained mission profiles.
- Launch schedule inflexibility.
- Weather dependency of launch and recovery.
- Little or no space rescue capability.

These characteristics and limitations contrast sharply with the autonomy, flexibility, maneuverability, responsiveness, survivability, and cost effectiveness required of military operations as lessons of experience and established in military doctrine.

Further, manned space vehicle programs and concepts have fostered the commonly held perception that the economics, technology, and safety of man in space will force the continuation of these nonmilitary characteristics into the future.

The National Command Authority and the Department of Defense rely heavily on unmanned satellites as vital elements in command, control, communications, intelligence, surveillance, reconnaissance, and warning. Unmanned satellites have additional problems relative to manned vehicles, such as inherent vulnerability to antisatellites, single mission utility, and the inability to adapt or to think. Therefore, the use of unmanned satellites compounds the problem by limiting the reliability of their support to military space vehicles and adding the problem of manned vehicles protecting, supplementing, or standing in for satellites. Balance and mutual support must be achieved between the manned and the unmanned military space systems. The manned vehicle must be capable of going "where the action is," including where the satellites are and can be in peacetime, conflict, and war.

THE NEED

In space, the need is to provide the military man with a highly cost effective vehicle system with the required military characteristics and capabilities that will secure the High Frontier to: (1) protect the United States resources from threats in and from space, (2) conduct needed aerospace offensive and defensive operations to use and protect the use of space by the United States and its allies, (3) enhance the land, sea, and air forces, (4) serve as a practical utility vehicle in the support of space assets and the exploitation of space, and (5) support as many aspects of U.S. national policy as possible.

THE SOLUTION

Studies to date have shown that limitations exemplified in the problem statement are not inherent in the new type of piloted military space vehicle termed the high performance spaceplane. The HPSP would perform missions throughout cislunar space and in the upper atmosphere comprehensively and do so with truly military characteristics and capabilities.

DESCRIPTION OF HPSP

To obtain high endoatmospheric performance, the initial HPSP is designed in the well-understood conical shape of the ballistic missile reentry vehicle. It is sized to carry one pilot, and its gross weight is approximately 6,000 pounds.

The extremely swept-wing conical reentry vehicle is approximately 23 feet long and has a base diameter of 52 inches. A storable propellant propulsion system is standard in the basic vehicle to provide propulsion for missions in low to medium altitude orbits. External tankage can be attached for increased velocity. The internal propulsion system is analogous to the range extension provided conventional aircraft by wing or belly tanks. The engine comprises a circumferential ring of small thrusters around the aft end of the vehicle. This nonconventional engine is called a plug-cluster (PCE) and is capable of operating efficiently at all altitudes from sea level to the vacuum of space. Individual thruster control provides thrust magnitude control and thrust vector control for steering.

The entire spaceplane is covered with ablative material covering lightweight tiles. This insulation provides maximum reentry thermal protection and permits distribution of the heat load by rolling the vehicle. It is expected that the airframe will be made of a nonmetal, composite material for its low weight, compatability with insulation tiles, and low cost.

During space flight and launch the pilot can sit upright. During endoatmospheric maneuvers, the pilot is seated in a partially reclined position, and a hatch replaces the canopy to restore the conical shape and provide protection from reentry heating.

Recovery is by the modern controllable parachute which permits piloted or automatically controlled flight to a landing at a small, unprepared site, an airfield, an aircraft carrier, a helicopter flight deck on a ship's fantail, etc.

For very high velocity change missions, such as rendezvous with a sequence of satellites or the placement of payloads in geostationary orbit, an external propulsion module or stage can be attached. Such a propulsion module attached to the aft end of the spaceplace would contain an RL-10 engine and cryogenic (liquid hydrogen and liquid oxygen) propellant tanks. The RL-10 is the engine used most in space and is well suited to the HPSP. The circumferential overlapping of the aft end of the spaceplane by the propellant tanks results in a lower overall length than would be obtained with a conventional cylindrical propulsion stage with the spaceplane as its payload.

An alternative cryogenic stage would be a version of the Centaur upper stage. If length is not critical, the Centaur type of upper stage can be used. The Centaur also uses the RL-10 engine.

OMNIMISSION CAPABILITY

The HPSP would provide a high degree of universality because it has:

- Configuration and performance that exploits both the space and atmospheric environment.
- Man's unique on-site capabilities.
- Man-machine unification in the vehicle.
- An unmanned operation mode.

Mission categories are exemplified by reconnaissance and surveillance; inspection and verification; antisatellite and anti-antisatellite; placement, supplementing, and standing in for unmanned satellites; on-orbit service, repair and update of satellites; and missions requiring multiple atmospheric entry and exit. The HPSP transforms the Space Shuttle Orbiter into an aircraft carrier in space and extends its military operations throughout cislunar space. Compatible with ground and air launch, the HPSP can also operate completely independent of ground operations and the Shuttle system.

CONFIGURATION REASONING

Let us consider the logic that dictates the specific configuration of the HPSP. Derived from the fundamental principles of rocketry, orbital mechanics, aerothermodynamics, and hypersonic flight, it is unlikely that the general configuration should change appreciably in the foreseeable future.

WHY CONFIGURED FOR ENTRY?

Entry capability is required for autonomous operation, proper energy management, and safety. Autonomous entry and recovery enables the HPSP to operate independently of recovery by the Shuttle Orbiter. Proper energy management is vital to mission performance. Safety is vital to mission success, and to the pilot.

In terms of energy management, the ability to enter and maneuver in the atmosphere empowers the vehicle with these important capabilities:

- Mission range extension by maximizing the propulsive velocity available to do mission tasks when less velocity is required to reach the atmosphere than to return to the Orbiter or other rendezvous point.
- Aerobraking at perigee in the atmosphere rather than requiring retropropulsion with its resultant weight penalty and loss of subsequent maneuver velocity.
- Use of aerodynamic lift to change the direction of flight (orbital plane change) and then to return to space flight. This energy-efficient maneuver is called the synergistic plane change and is efficient for a vehicle with the lift-to-drag ratio and low drag of the conical configuration of the HPSP.
- Use of aerodynamics to maneuver to a landing point on Earth and to minimize pre-entry propulsive maneuvers.

In safety terms, the entry capability enables a recovery return either to Earth or the space station such as the Orbiter, depending on the time available to reach sanctuary, the specific failure problem or damage that forced the premature recovery or abort, medical needs, or docking risks to the Orbiter. With sidecars it can carry a number of passengers to and from the Orbiter, another manned vehicle, or a satellite.

Without the entry and landing capability, a manned orbital transfer vehicle or other manned vehicle is not efficient, safe, or truly military.

WHY THE CONICAL SHAPE?

The cone is the most understood and tested shape for reentry. It is the shape of the ballistic missile reentry body for the same reasons as the HPSP, particularly the need for low drag and high lift-to-drag ratio. These result in the minimum loss of velocity during endoatmospheric maneuvers. Therefore, the least amount of propellent is consumed in returning to space and the maximum footprint or area in which the vehicle can fly is obtained. The cone presents the smallest surface area consistent with high aerodynamic performance, and surface area means weight in the thermally protected reentry body.

Minimizing vehicle weight is critical to propulsive maneuverability in space, to maximize vehicle payload capability, and to performance and size of any launch vehicle, whether it is the Orbiter, an expendable rocket, or a launch aircraft.

The conical shape is correct for the generic, highly maneuverable spaceplane. Orbiter vehicles designed to meet a substantial internal payload volume specification may require the winged, nonaxisymmetric shapes exemplified by the Orbiter but will be penalized in performance. Because there is no drag in the vacuum of space, payloads can be carried externally and the size and weight of the HPSP is therefore minimized, resulting in optimal payload maneuverability performance.

Returning to the analogy of the ballistic missile, we observe that each missile pushes its payload which is either the next stage or the missile payload, externally mounted for the best overall performance.

We may conclude that the conical reentry body is used best for vehicles with specifications for maximum payload maneuverability, maneuverability with small internal payloads, synergistic plane changes, lightest weight, compatibility with launch by the Orbiter or the MX ICBM booster, near term availability, and lowest cost. Other shapes may be best if internal payload volume is the driving requirement.

The latter type of vehicle is exemplified by the Orbiter. Vehicles of this type may be characterized as logistic vehicles. That is, they are principally launch vehicles or payload recovery vehicles for operation in the lower orbits. Their use in higher orbits or for high velocity change maneuvers is not cost effective.

AIRCRAFT LAUNCH

Analysis of aircraft launch for the HPSP has shown that the combination of the Boeing 747-200F freighter aircraft and a rocket launch vehicle or booster would place greater than 20,000 pounds into low orbit. Therefore, the HPSP with a large payload and amount of propellant in HPSP external tanks could be orbited. Alternatively, two or three HPSPs could be orbited by one launch vehicle. Aircraft launch would be a flexible, military means for delivering large payloads to orbit where they could be transferred to orbits by the HPSP.

The launch vehicle is of conventional design. The Titan LR-87 type engine would power each of two strap-on boosters attached to a two stage core rocket powered by a Titan LR-91 engine on the first stage and three Pratt & Whitney RL-10 engines on the second (final) core stage. Liquid oxygen and commercial liquid propane would be the propellants for the Titan engines. Ease of handling, high performance, and low cost are obtained with this launch vehicle.

With the availability of the MX booster and Orbiter as launch vehicles for the HPSP and the availability of rocket engines for aircraft launch, the next step is to obtain the cislunar maneuverability, payload maneuverability, and omnimission capability of the small HPSP uniquely. The vehicle can gain and protect the "high ground" in the High Frontier at a comparatively minor cost of acquisition.

PRIORITY

The aerospace forces of the United States must be able to go immediately to where its satellites are and where the threat is. As resources permit, the logistic-type vehicle might then be developed to operate under the protection of the high performance spaceplane.

APPENDIX E: ADVANCED TECHNOLOGY SPACEBORNE DEFENSE SYSTEMS

The High Frontier concept of layered defense visualizes two layers in space. The first deployed layer attacks ballistic missiles in the early stages of trajectory and provides some capability for denial access to space of other hostile objects. The second layer, utilizing more advanced technology, would be capable of engaging the more difficult targets of individual reentry vehicles in mid-trajectory and establishing strong capabilities against all hostile objects in near Earth orbits.

These concepts of spaceborne defenses tend to evoke images of Earth satellites armed with directed energy (beam) weapons, shooting at enemy missiles and at each other. Such images may become reality in the future, but while significant beam weapon capabilities have been demonstrated in the laboratory, their deployment in global defensive systems is too far in the future to meet the urgencies of the High Frontier study. Nevertheless, the ultimate promises of beam systems, especially in the light of intensive Soviet efforts to create such weapons, demand that a well planned and funded U.S. R&D program, at least of coping proportions, be conducted. Technological breakthroughs in beam weaponry might well tip the strategic balance at any time in the future. Still, when faced with near term defensive needs, it is prudent not to count on technological breakthroughs.

It is important to note that all requirements for the High Frontier layered defense concept can be met without the prior development of beam weapons or any other technological breakthrough.

BEAM WEAPONS

Directed energy technology involves the generation of energy and its precise and nearly instantaneous delivery to objects of interest at the long ranges required for space operations. Directed energy technology also provides opportunities for the United States to pursue valuable industrial and commercial operations in space.

Directed energy technology encompasses a family of generically similar concepts that collectively contain potential for major breakthroughs in military capability. The systems generate and project intense electromagnetic energy (radio frequency, optical, X-ray) or subatomic/atomic particles (electrons, protons, ions) to perform a variety of missions including target tracking and destruction, electronic warfare, and surveillance. These include:

- High energy lasers (HEL)
- Particle beams (PB)
- High power microwaves (HPM)
- Electromagnetic pulse (EMP)

Common characteristics of these three technologies are the high propagation velocity of the energy and the ability to focus that energy into a very narrow beam (from which stems the popular term "beam weapons"). Other characteristics are:

- Effectiveness can be achieved at extremely long ranges.
- Target evasive maneuvers are difficult.
- The weapon can engage several targets within a short time span.
- The weapon usually possesses multishot capability.

Directed energy weapons operate on different physical principles than conventional weapons systems. Lasers, particle beams, and microwaves all project energy at (or near) the speed of light, i.e., 186,000 miles per second, but the beam ranges and the modes of interaction with both the targets and the intervening environment differ greatly.

Performance depends on a complex relationship between a number of system, target, and engagement parameters. For example, lasers might kill a given target by depositing large amounts of energy upon its surface; microwaves might induce effects penetrating some distance into the target; and particle beams might penetrate deeper still. The concomitant kill mechanisms might be quite different in the three cases, and the total energy required for the kill might vary considerably depending upon the mechanism. On the other hand, the propagation losses from the weapon location to the target might also vary greatly with the choice of the weapon device, possibly making the easiest kill mechanism the hardest to achieve from a given distance.

For lasers and microwaves, the choices of wavelength and transmitter diameter determine most of the propagation characteristics. In addition to the propagation characteristics, the question of atmospheric effects must also be considered. Generally, a laser beam is more strongly affected by inclement weather and dust (i.e., aerosols) than microwaves are. Atmospheric turbulence (like the shimmering of the air seen over a hot surface) also perturbs laser beams, requiring clever ''adaptive optics'' systems to correct the disturbances. Moreover, there will be some combination of wavelength, beam diameter, and power level where the atmosphere will break down (like a spark), with great loss of power from either the laser or microwave beam. The ability to achieve energy levels sufficient to kill a target varies greatly among the various choices. In addition to the factors already discussed, the nature of the hardware and its overall efficiency must be considered.

Particle beam weapons using electrons within the atmosphere or neutralized ions in space are attractive. They have two potential advantages over electromagnetic beams, i.e., lasers and microwaves; namely (1) they are probably immune to weather and dust clouds when used within the atmosphere, and (2) they penetrate deep into the target making countermeasures difficult. On the other hand, the required hardware may be very cumbersome and the range may be so short (in the atmosphere) that their use in strategic defenses is not feasible.

From the foregoing, it should be clear that the key to militarily useful beam weaponry is the ability to deposit lethal radiant flux density on the target at the maximum possible range. The levels of lethality may vary by many orders of magnitude according to the nature of the kill mechanism. It is also important to note that all directed energy weapons have a long range ''soft'' kill potential in addition to a shorter range ''hard'' kill potential. That is, they can burn out or temporarily disable target electronics and/or electro-optical systems at ranges beyond the ''hard'' kill range. The range and effectiveness of such kills vary widely depending on the target and specific weapon and engagement scenario.

Directed energy technology has great potential for both military and civilian applications. Used in space, beam weapons may offer opportunities for worldwide projection of military force with essentially instantaneous destructive capability against satellites, aircraft, cruise missiles, intercontinental ballistic missile and submarine launched ballistic missile boosters, reentry vehicle buses, and surface targets. Land based systems, particularly high energy lasers with space based mirror relay as well as high power microwave devices, could also provide early capabilities against some of these targets. The impact of the development of such systems would be immense, definitely affecting the balance of world power. In addition, there are potentially important civilian applications, such as power transfer and propulsion, where directed energy technology, in the longer term, could have important economic benefits.

The potential mission which drives system requirements in beam weaponry the hardest is ballistic missile defense (BMD), against which 1,000 or more ICBMs may be launched simultaneously. This mission will stress most of the technology requirements. The BMD application has the highest payoff during the boost phase of missile trajectory when the number of targets is minimum and their vulnerability is maximum. The destruction of the carrying missile negates all of its multiple warheads. This phase lasts about 240 seconds, the burn time for the missile booster.

Another important potential application of beam weapons is destruction of high flying aircraft. Aircraft at high altitude, i.e., above most of the diffusing layers of atmosphere, can be destroyed by spaceborne lasers of selected wavelengths.

Finally, practical potential applications for beam weapons include negating satellites (ASAT), or for the defense of satellites (DSAT). The major advantage of beam weapons would be that they could very rapidly destroy Soviet satellites or ASAT systems.

Realizing the great potential of beam weapons depends in large measure on excellent command, control, and communications (C^3). Elements of command and control that must be considered in the development of beam weapons include operational control, tasking, weapon status monitoring, and battle damage assessment. Establishment of a coherent, nonambiguous space track picture of the surveillance volume is critical to effective beam weapons operation. The global deployment of directed weapon systems will also require an extensive communications support capability that must be survivable and enduring throughout all phases of conflict.

For the near term, the only directed energy weapons potentially useful for space applications are ground based laser and microwave systems in an ASAT role. Either type can probably be deployed at a single, preferably elevated, ground site. The technical risk appears to be low for the microwave system but medium to high for the laser system, depending on the target range of interest. For example, the laser system risk is medium for target ranges that include the present Soviet low altitude ocean reconnaissance satellites, but high for longer ranges because sophisticated adaptive optics would be required. The microwave system suffers at present from a lack of knowledge and agreement on the lethality of the kill mechanism. If the microwave power levels proposed are determined to be lethal, this system may be the better choice.

In determining the appropriate technical options for beam weapons, U.S. decisionmakers must evaluate the advantages and disadvantages of the diverse directed energy weapon technologies in a variety of deployment configurations.

Particle beams, while highly attractive, are at a very early stage of development. High power microwave technology has only recently started to receive attention as a potential beam weapon, although its relative familiarity and potentially less demanding system requirements may result in its being used very early if the lethality uncertainties are favorably resolved. Thus far, laser system technology has received the most attention and is the most advanced. The several technologies which show promise are described below.

HIGH ENERGY LASERS (HELs)

Low and moderate energy lasers have made important contributions in numerous applications, including medicine, science, cartography, communications, range finding, and target designation. Potential applications for high energy lasers involve not only weapons which can damage targets, but also laser radar, laser generation of fusion power, materials working (welding, heat treatment, etc.), and laser isotope separation.

The key components of a hypothetical laser weapon system include both the laser itself, which generates the high power light, and the beam control subsystem, which aims the laser beam at the target and focuses it on a vulnerable spot on the target. Like other weapons, the laser weapon system must also have a fire control subsystem which acquires all the targets that need to be engaged, selects the one to engage, and tells the beam control subsystem where to look to find it. Finally, the fire control subsystem decides when the target has been destroyed and designates the next target.

Shortly after the invention which made the generation of high energy laser beams possible, it became apparent that a laser damage weapon system, if it could be developed, would have some particularly attractive features. For example, since light travels at a speed of 186,000 miles per second, the lethal flux would arrive at the target almost instantaneously, and there would be no requirement to "lead" the target. It takes six-millionths of a second for laser light to travel one mile, and in that time a supersonic airplane travelling at twice the speed of sound will travel only a little more than one-eighth inch. Because of its pinpoint rather than area effects, a laser weapon could be used to selectively attack and destroy single enemy targets in the midst of a host of friendly vehicles.

A laser weapon can be expected to handle a large number of targets, even if the targets are coming from all directions. For each "shot" the laser takes, relatively small amounts of fuels are used to generate the beam. Thus, there is the potential for storing a large number of shots per installation (or a large magazine per weapon). Finally, since the beam is steered by using mirrors, the laser weapon has the potential to move rapidly from target to target over a wide field of view.

Although such a system has enormous potential, development efforts must also address those characteristics of high energy laser weapons which tend to mitigate the promise of such weapons. For example, a successful laser engagement occurs only when the beam burns through the target surface and destroys a vital component (e.g., the guidance system) or ignites a fuel or warhead. Thus, while the energy is delivered instantaneously, the laser beam must dwell on the target to destroy it. Furthermore, "jitter" of the focused spot over the target smears the energy in the beam over a larger effective spot size, increasing the time required to damage the target. Thus the beam control subsystem must hold the beam steady on the target aimpoint. To do this, the target tracking and beam pointing functions of the beam control subsystem must be exceptionally accurate.

Fire control for laser weapons will have to be especially capable. It must be able to recognize and classify a host of potential targets, and determine which to engage first. In addition, to realize the firepower potential of a laser weapon, the fire control must be quick to recognize that the target being engaged has been damaged sufficiently that it can no longer perform its mission, so that the laser beam can then be moved to the next target.

A final example of a characteristic which tends to mitigate laser capabilities is the effect the atmosphere has on the laser beam. As a function of the wavelength of the laser energy, the atmosphere absorbs some of the energy being propagated, causes the beam to "bloom" or defocus, and adds jitter to the beam. Interactions between the high power beam and the atmosphere effectively increase the spot size on target, lowering the peak intensity and increasing the necessary dwell time.

Since in the vacuum of space the laser beam does not have to contend with the degradations caused by the atmosphere, space has often been referred to as the "natural" environment for laser weapons. In this vacuum one can envision achieving the very long weapon ranges of operation needed to contend with the vast volume of near Earth space. In addition, at long ranges, the stressing requirement to point accurately is ameliorated by the demand for only low angular tracking rates. Thus, it may be easier to hold the beam on targets at the high velocities typical of objects in near space (5-10 kilometers per second).

The high energy laser scientist can envision a weapon in a high density threat environment that methodically moves from target to target over its all azimuth coverage, focuses the beam on the target, holds the selected aimpoint despite the target's speed and maneuver, burns through the target skin, and destroys a vital component or ignites the fuel or warhead. Then, with instructions from its sophisticated fire control system, the weapon switches the beam to the next target providing greatest threat and so continues through a number of successful engagements before the fuel is expended.

Although many different lasers were discovered in the 1960s, none was suitable for high energy applications; some additional discoveries and inventions were needed. A principal discovery was that gaseous molecular lasers were possible. This discovery led to efficient lasers which generated their energy in the infrared portion of the spectrum. The next step was to invent a way to generate the energy required to operate the laser in an efficient and scalable manner. The required invention was made in 1967. It was the carbon dioxide gas dynamic laser, or CO_2 GDL for short. The CO_2 GDL was the first flowing gas phase laser that appeared to be scalable to very high energies, and as such paved the way for serious consideration of a laser damage weapon system. In recent years, other high power laser concepts have been developed on this same basic principle, i.e., flowing gas, including the electric discharge and chemical lasers. Using these concepts, high energies have been generated at differing wavelengths.

The U.S. DOD HEL program involves development of many technologies and is truly multidisciplinary. In addition to the usual scientific and engineering activities, special attention is devoted to the understanding of how a laser beam propagates through the air and interacts with the target. Moreover, in view of the possibility that the potential enemies may eventually develop a laser weapon system, significant resources should be devoted to an investigation of techniques by which systems can be hardened to increase their survivability in a laser weapon environment.

The required operating time of the laser is important as it places demands upon the power source. If the laser is chemically fueled, then an adequate amount of fuel must be carried into space. Alternatively, if the laser is primarily electrically powered, then a potentially large, heavy power generating, storage, and conditioning system may be required, or possibly a small nuclear power system. Soviet emphasis on small nuclear power systems in space contrasts with an essentially nonexistent U.S. program.

The highest power currently demonstrated chemical laser system is sufficiently capable to support early development of a land based system for a limited ASAT capability. Somewhat higher brightness systems would increase the utility of ground based ASAT systems and when configured for space based operation would easily handle the ASAT requirements and begin to handle more demanding engagements.

A space based chemical laser force could provide for rapid global projection of U.S. power in conflicts of limited nature. It could provide simultaneous continental U.S. and fleet air defense and could be used to attack airlift lines of supply and airborne warning and control aircraft. This ability is considered unique since no other system has potentially instantaneous global antiaircraft coverage. Space based chemical lasers of exceptional brightness are technologically feasible in that no insurmountable technical issues have been identified and multiple potential solutions are possible for known critical issues. Deuterium fluoride chemical lasers have demonstrated the highest average power to date, although none of the chemical laser programs are specifically designed to demonstrate space laser technology. Current efforts are aimed at developing new chemical laser configurations designed to scale to much higher power levels.

Another promising technological choice is the free electron laser which allows tunable wavelengths and promises high energy conversion efficiencies. The current free electron laser experiments are intended to verify the analytic models and to demonstrate efficient operation using linear accelerators. These experiments will be continued to increase the efficiency to 30 percent. Further plans include wavelength scaling into the visible spectrum with higher energy linear accelerators, linear accelerator stability experiments, and initiation of a moderate average power free electron laser to be completed in the mid 1980s.

Another possible technology which could lead to an effective beam weapon is X-ray laser. In space, much of the energy of a nuclear explosion is released in the form of X-rays, which are extremely short wavelength radiation. A 300-megaton blast in space will deliver satellite damaging energy at a distance of several hundred miles. Thus, without any special techniques, a very large yield burst in space might incapacitate a space vehicle. The USSR exploded a 60-megaton device in the 1960s and has boasted that the yield could be increased, but there is no evidence of such large weapons in the Soviet inventory today.

The abundant X-rays from an extremely high altitude burst, i.e. in space, attenuate with distance. If a means can be found, perhaps through laser techniques, to focus even a small fraction of the X-ray energy of a nuclear device in space, then the directed beam would constitute a weapon against objects in space. A shielded object may be able to survive one X-ray "shot" but would probably succumb to a second shot.

Such an X-ray laser system would, of necessity, destroy itself to operate. It would have to destroy itself even to defend itself against a single kill vehicle. This creates an inherent self-contradiction for such weapon systems if deployed in space. Technologists reply that a single shot might be designed to generate more than one X-ray beam, so that at least tens of targets might be hit. But imposing simultaneous multiple hit requirements, with the attendant difficulties of fire control, would appear to impose such additional engineering difficulties that X-ray laser systems should not be counted on for essentially near term defensive imperatives.

PARTICLE BEAMS

A particle beam is a stream of highly energetic atomic or subatomic size particles such as electrons, protons, hydrogen atoms, or ions. (By comparison, laser beams are composed of radiant energy photons.) An electron beam would resemble a lightning bolt. Presently, aside from potential applications as weapons, particle beam machines have potential for use in inertial confinement fusion for energy generation, nuclear weapons simulation, heating and welding, high intensity microwave generation, geophysical investigations, energy transmission, medical treatment (e.g., cancer), and basic physics experiments.

There are three key components of a hypothetical particle beam weapon system. First, there is the source of the beam—the beam generator—consisting of a particle accelerator and its associated supply of electrical power, energy storage, and conditioning. The accelerators are similar to those used in research in elementary particle physics except that currents in the beam are much higher. Second, there is a beam control subsystem to aim the beam at the target and determine that the beam has hit. Last, the particle beam weapon must have a fire control subsystem which acquires all the targets that need to be engaged, selects the one to engage, and tells the beam control subsystem where to look to find it. Then the fire control system decides when the target has been destroyed and designates the next target. These fire control functions do not differ materially from those of fire control subsystems for other more familiar weapons.

An appreciation for the damaging effect of highly energetic particles striking an object can be gained by recognizing the damage lightning can do when it strikes a tree or a house. (As a matter of fact, since the beam resembles a lightning bolt, technologists will often refer to a "shot" from a particle beam accelerator as a "bolt.") In high energy physics, experimenters have long been aware of the ability of the highly energetic particles produced by atom smashers to penetrate into materials. As the beam penetrates, it transfers some of its kinetic energy from the particles to the material and, in addition, generates secondary radiation in the material, which can also disable the target electronics. If there are enough particles in the beam hitting the target, the rapid transfer of energy to the material cannot be dissipated by the material. Thus, the beam can cause a hole to be burned or melted into the material, or a fracture from thermal stresses as a result of the rapid deposition of energy. Another example of effects can be taken from discoveries in the early days of space flight. Energetic charged particles generated largely by the Sun are trapped in the Earth's magnetic field, thereby forming the "Van Allen" belts. These natural particle beams require spacecraft designers to build shielded and resistant satellites if flights in or through these belts are to occur without damage to such "soft" components as computers or electronics.

Thus, one can envision a weapon based on a stream of highly energetic particles that travel at nearly the speed of light. This stream of particles would penetrate the metal skin of the target, transferring a large fraction of the energy in the beam to the target. Initially, as the beam enters the target it would damage electronic components and, as the beam continues to deliver energy to the target, ignite fuels and explosives and/or create holes in the target.

Particle beam technology is in the early research and exploratory development phases with fundamental issues of feasibility to be resolved. The next major milestone in the program is establishment of scientific feasibility by addressing the key physics and technology issues that, once resolved, will indicate whether particle beam weapons are practicable.

A particle beam weapon is a system which produces a high energy, small diameter beam of either neutral atoms or charged ions to disable the target. The choice of charged versus neutral particle beams in the design of weapons depends on the deployment mode. Charged particle beams can only be used in the Earth's atmosphere at relatively short ranges. Neutral particle beams can only be used in space where the range limiting effects of the atmosphere are not present and long range engagements are possible.

For many reasons, neutral particle beam range requirements generally are on the order of 10,000 kilometers. If ranges of 10,000 kilometers are practical, then the system is very attractive since a relatively small number of neutral beam platforms could meet a large number of space defense requirements.

HIGH POWER MICROWAVE (HPM)

The term "microwave" as used here encompasses the frequency regime just above conventional radar frequencies to just below most laser frequencies. This is a loose definition, and there is some overlap at the boundaries of the defined regions.

High power microwave weapons concepts are based on a number of emerging high power source technologies which may lead to substantial improvements in radars, communications, etc., as well as to the possibility of utilizing microwave radiation in weapons. Recent advances by both the Soviet Union and the United States indicate that orders of magnitude increases in averaged and pulsed power output at millimeter and higher wavelengths are now possible.

In the past, the frequency domain between lasers and high frequency radars (that is, between micron and centimeter wavelengths) have been unavailable for widespread military exploitation. Although there are some millimeter and submillimeter wave sources, these sources are limited to very low powers. Newer technologies, generally based on relativistic electron beam approaches, can circumvent the power limitations of the earlier devices and make available for the first time substantial powers in the millimeter and submillimeter regimes.

The key interest in the directed energy technology with high power high frequency microwave systems is the potential for either destruction of space vehicles or burn out of their electronic components at very long range. With the high powers now achievable microwave impulses are potentially lethal, particularly against targets such as cruise missiles, remotely piloted vehicles, aircraft, and possibly RVs. Satellites and other targets in space can also be killed by rapid heating of structural and functional components as well as by inducing currents which damage sensitive electronic components.

Like the Soviet Union, the U.S. is developing a high power, high frequency microwave capability. The Soviet results to date are quite impressive, and it is generally agreed by most U.S. researchers that the Soviets enjoy a several year lead in some technology areas.

The dimensions of microwave weapons systems that have been previously proposed are very large. More recent considerations of HPM technology indicate the possibility of smaller systems which could be space based. The technical challenges for this type system appear lower than for any other directed energy weapon. However, there is uncertainty and lack of agreement on the lethality of microwave energy at the power levels studied. Furthermore, the uncertainty is associated with details of the target design since a major kill mechanism is leakage energy getting into the electronics that causes the damage.

ELECTROMAGNETIC PULSE (EMP)

EMP is related to HPM weapons because of phenomenology, propagation, and effects similarities. While EMP is generally similar to radio waves, it exhibits important differences. EMP waves include a broader range of frequencies and amplitudes than radio transmitters can produce, and electric fields associated with EMP can be millions of times greater than those associated with radio waves.

All nuclear explosions generate an EMP, although the intensity, duration, and area over which the pulse is effective varies with the altitude of the burst. Unlike the relatively localized EMP effects experienced with surface bursts, high altitude detonations—those occurring 19 miles up or higher—blanket a line-of-sight radius on Earth. For a blast 50 miles up, the affected ground radius would be 900 miles. And for an explosion centered over the U.S. at an altitude of 200 miles, the entire continental United States (including parts of Canada and Mexico) would receive the EMP.

For high altitude bursts with yields of a few hundred kilotons or more, electric fieldstrength will vary by no more than a factor or two over most of the area showered by EMP. Maximum EMP can reach 50 kilovolts per meter. Virtually every electrical conductor will serve as an EMP antenna unless it is adequately shielded.

Unlike lightning, EMP imparts less energy but delivers it 100 times faster, usually faster than lightning arresters can handle. For a large high altitude nuclear burst, the fields radiated onto the Earth's surface peak in 10 nanoseconds—roughly 100 times faster than lightning. This fast rise time represents a double edge sword. First, it means the spectral energy will be distributed much more broadly throughout the electromagnetic band—including the lower microwave range. Second, the rise time is so rapid that an EMP can travel through a system—destroying sensitive electronics (similar to HPMs effects described in the previous sections) along the way—before lightning arresters or other defensive power-shunting switches can respond to the surge.

For these reasons, EMP is different from any other electromagnetic environment usually encountered that protection practices and components for non-EMP environments—radio frequency interference, lightning, radar, etc.—are not directly applicable to EMP problems. For systems whose continuous operation is deemed critical, such as military surveillance, communications, and attack units, EMP protection—known in the jargon as "hardening"—becomes essential. Hardening our communications, command, control, and intelligence systems against electromagnetic pulse will be one of the major strategic undertakings of the 1980s.

There is a striking contrast between civil and military approaches to coping with potential EMP disruptions. While the nearly universal military approach has been to harden systems of interest, this is not a feasible civil measure. Military attack and communication systems cannot afford to shut down even momentarily during attack periods, whereas civil preparedness systems can afford to be out of action for periods running from minutes to days. So, while some attempt has been made to harden civil systems such as the Emergency Broadcast Network, another common strategy has been to analyze likely damage should an EMP occur and then to develop contingency plans to cope.

However, the electric power industry has been too complacent about the threat to its potential vulnerability to take even these measures, and while the military has aggressively sought to EMP-harden its most important facilities and weapons over the past 15 years, it is quite dependent on several civil systems that appear potentially quite vulnerable to EMP—notably, the nation's electric power and telecommunications industries. In the event of war, these military dependencies on non-EMP-hardened networks could prove an Achilles heel to national defense.

In thinking about focused energy sources, it is tempting to prognosticate that other electromagnetic energies might by some means be self-generated in space. If it were possible to create electromagnetic pulses in space, not having available the atmospheric modes which contribute to EMP formation at lower altitudes, a hypothetical space EMP weapon could be formidable as a weapon against space systems over areas of hundreds of square miles. However, "sweeping" space in such a manner would also disable friendly space systems.

Because we cannot discount other nations developing advanced antisatellite kill mechanisms employing a broad spectrum of electromagnetic options, ranging from EMP through the optical bands to X-rays, it is essential that conceptual and development work on such sytems be initiated and continued apace by the U.S. and its defense allies.

APPENDIX F: CIVIL DEFENSE

THE NEED

Under the policy of Mutual Assured Destruction (MAD), this nation's citizens remain unprotected hostages to the Soviet Union's steadily growing nuclear strike force. By contrast, the Soviet Union has taken substantial steps to preserve the lives of its population in a nuclear war.

The protection of our citizens must be prime, but civil defense is also a key element of the strategic balance. An effective civil defense would reduce the possibility that the U.S. could be coerced in a time of crisis. A space based anti-missile defense would go a long way toward providing a defense of cities and critical installations. But until a global missile defense system is achieved, greater priority should be given conventional civil defense measures.

As early as the 1950s, the Soviets had an extensive civil defense program which included fallout shelters, planned evacuation of cities, storage of grain and other foodstuffs, and required survival training for every citizen of the USSR. In 1961, civil defense was transferred from civilian to military control under Marshal V.E. Chuikov. In 1972, civil defense was further elevated in overall Soviet strategic planning, and Colonel General A.T. Altunin was appointed commander of the civil defense program with greatly expanded responsibilities.

Civil defense is regarded by the Soviet Union as "a strategic factor" that will make "a major contribution toward victory" and was so stated in a policy paper, *The Philosophical Heritage of V.I. Lenin and Problems of Contemporary War,* edited by General Major A.S. Milovidov and translated in a U.S. Government Printing Office document in 1974.

Thus, the upgrading of civil defense to a strategic status, as recommended by High Frontier for the U.S., has already been an integral part of Soviet strategic planning for its own population and industry for many years. This in itself should be sufficient reason for the U.S. government to scrap the discredited policy of Mutual Assured Destruction which dictates, among other things, that American lives should not be protected through civil defense.

However, renunciation of MAD will not in itself reduce the number of lives at risk or improve our capabilities to deter Soviet aggressions. The dangers resulting from 15 years of unilateral U.S. adherence to the MAD strategy can be overcome only by protection of lives, property, and defensive forces. A viable civil defense program is a necessary complement of decoupling from MAD.

In the next few years, before active defense weapons can be perfected and deployed, American lives and essential recovery assets can be protected best by rapidly creating a strategically significant, yet relatively low cost, civil defense system—the first step toward Assured Survival. Assured Survival can be attained if the United States, as soon as possible, deploys both an effective civil defense system and military defensive systems capable of actively protecting America and its population.

ESSENTIAL REQUIREMENTS

The primary requirement of any strategically significant U.S. civil defense system is credibility: its credibility to the Soviet Union and other hostile nations, to the majority of Americans, and to the allied and unaligned countries. The credibility of our civil defense preparations will depend on their widely recognized ability to save a great number of lives during both an attack and the following recovery period, and on the ability to protect enough essentials of our agricultural and industrial assets to give reasonable hope for the recovery of national power and our modern standard of living. The continuing support of the majority of citizens requires that civil defense preparations also must be widely recognized as making an important contribution to the prevention of wars. American leaders must stress the positive, hopeful advantages of serious survival preparations.

PRESENT STATUS

Today, U.S. civil defense capabilities are dangerously inadequate, and most of the few that do exist are becoming increasingly ineffective. Outdated and even potentially life endangering survival instructions continue to be stockpiled for crisis distribution. Congress has failed to appropriate enough money to make a planned, nationwide, modest start of crisis relocation planning (CRP), which is the organized and supported evacuation of the populace from probably targeted areas.

The limited funds made available for a few CRP area tests continue to be partially diverted to less disturbing, more acceptable emergencies. The probable fiscal year (FY) 1982 civil defense funding of about $128 million will amount to approximately 59 cents annually per capita. Most of this annual pittance will be spent on measures to mitigate the dangers from floods, windstorms, and other natural disasters. In purchasing power, the probable $128 million FY 1982 funding will be less than one-sixth of the 1962 civil defense budget.

The responsibility for civil defense now rests with the Federal Emergency Management Agency (FEMA), the catch-all disaster and hazard reduction agency created by President Carter. FEMA lacks the necessary close association with the Department of Defense, the ability to concentrate on war survival problems, and the extremely able executives and engineers essential for expediting the deployment of a strategically significant civil defense system. Currently, FEMA is characterized by indecision and by avoidance of many of the hard problems and inevitable controversies of realistic nuclear civil defense.

MEASURES TO ATTAIN A STRATEGICALLY SIGNIFICANT CIVIL DEFENSE SYSTEM

The initially low cost civil defense system outlined in this paper is based upon an enhanced civil defense program advanced during the Carter Administration to implement the policies of Presidential Directive 41, "U.S. Civil Defense Policy," dated September 29, 1978. It is called Program "D." This much studied option relies for its lifesaving potential primarily on crisis relocation planning and includes operational systems, training, shelter equipment, and seven million additional sets of radiation monitoring instruments.

CRP should enable some 80 percent of Americans to survive a large scale attack in the mid-1980s, provided that (1) the majority of the risk area population will have been evacuated to host areas prior to the attack and, (2) fallout protection will have been developed and other crisis actions completed.

It includes improved warning, and sheltering capabilities for the in-place protection of Americans, should time and circumstances preclude crisis evacuation. CRP already commands strong bipartisan congressional support; Program "D" is the most effective option being seriously considered.

The March 30, 1981, FEMA paper, "Enhanced Civil Defense Program to Implement PD 41 Policies," describes in considerable detail the objectives, measures, and estimated costs of Program "D." This five-year program is designed to carry out objectives of the new Title V of the Federal Civil Defense Act of 1950. The FEMA paper also describes Program "D-Prime," a seven-year, stretched program with costs largely deferred until its last three years.

Rather than stretching the minimal Program D to seven years for implementation, High Frontier recommends a compression to completion within three years. To help counter the worsening dangers from Soviet nuclear weapons and transform Program D into a more effective and credible civil defense system, the following nine changes and additions should be incorporated:

1. The President must give civil defense his strong and continuing support. Furthermore, over the years the President must direct responsible civil defense organizations to balance funds and manpower allocated for measures to reduce war dangers against those covering peacetime emergencies. Most Americans expect their President, members of Congress, and appointed officials to provide for the common defense—including most certainly the defense and preservation of the lives of our citizens. Leaders who advocate and support realistic civil defense can win majority support.
2. Civil defense responsibilities (including the mitigation of life endangering effects of natural and manmade peacetime disasters) should be removed as soon as practical from the FEMA and reassigned either to the Department of Defense and/or the White House. The strategic importance of realistic civil defense—including its contributions to reducing casualties and other losses, to strengthening deterrence, and to improving the prospects for recovery—necessitates its being coupled with the Department of Defense and funded accordingly. In addition, able executives and engineers—the kind of men and women who always can get excellent jobs—are much more likely to accept civil defense work under DOD than with FEMA. The President should appoint a high level commission to determine within a few months the best way to restructure and realign a new and much more effective civil defense organization with other agencies, such as the Department of Defense.
3. "Self-help civil defense"* should be strongly advocated and taught by our official civil defense organizations during the years when Americans inevitably will continue to have far fewer blast shelters, smaller food reserves stored outside probable targeted areas, and generally less effective civil defense and active defense systems than the Soviets have developed in the past 20 years. Self-help civil defense, or "stop-gap civil defense," can go a long way toward filling the gap between our present ineffective civil defense and the highly effective civil defense system we should have and could have several years from now. House Resolution 7032 provides an example. It is a bill to amend the Federal Civil Defense Act of 1950. This bill was enacted in September 1980 as Title V of the Federal Civil Defense Act. Title V specifies that " . . . the President shall develop and implement a civil defense program which includes" . . . 14 program elements. One of these elements is

* An example of self-help instructions is the Oak Ridge National Laboratory report, *Nuclear War Survival Skills*. This book is a guide, largely based on realistic field tests with typical American families in several states, to help people unfamiliar with the effects of nuclear weapons improve their chances of surviving a nuclear war. Tens of thousands of privately reproduced copies of *Nuclear War Survival Skills* are currently being used to build fallout shelters and to make ventilating pumps, fallout meters, and other homemade life support equipment. If tens of millions of citizens could acquire basic survival knowhow via similar literature, then additional millions of lives could be saved.

"(12) the improvement of and training in self-help nuclear war survivor skills."

4. CRP should be greatly accelerated and essentially completed within three years of funding. A stretched out, less effective CRP program than Program "D," called "D-Prime," appeared to key civil defense supporters to be the best CRP program attainable. The stretched program was scheduled for completion seven years after an initial civil defense annual budget of $167 million was voted. The seven-year time schedule assumes that specified annual increases in funding will be forthcoming each year. (To date, Congress has failed to appropriate sufficient funds to make even a modest start of a stretched out program.) CRP for the next few years will be the most cost effective system to save lives and improve this nation's survivability. However, if the Soviets continue their present pace of constructing blast shelters in targeted areas and deploying both offensive and defensive weapons, within five years or so the Soviets may logically conclude that they can win a nuclear confrontation or war without ordering an evacuation. Then a national capability to evacuate high risk areas during a crisis, attained too late, is likely to be obsolescent. Unless a highly effective strategic defense is deployed, a great number of blast shelters will have to be built before the end of this decade if civil defense is to become an important addition to our defenses against the Soviet Union. Unfortunately, even excellent strategic defense systems are unlikely to ensure perfect protection. If urban Americans are thus actively protected, they will face less overwhelming threats and should be more supportive of fallout shelters and other passive-defense preparations to help them survive degraded attacks.
5. A major food reserve element should be added to Program "D." No civil defense program is meaningful without provision for emergency supplies of food. Strategic food reserves should be stockpiled in numerous facilities in the host areas. Austere basic foods—especially powdered milk for infants, wheat, and bulgur (partially debranned wheat that has been steam cooked and redried)—are essential for host area stocks. Provision of these basic foods will assure potential urban evacuees that they will not get painfully hungry if optimistic plans to redirect normal food deliveries to points in host areas do not work out. In an undisciplined society, an evacuation cannot be maintained if evacuees begin to suffer hunger. Our crisis evacuation, sheltering, and post attack recovery plans will be further improved and made credible if preparations are in hand to use local and trucked-in supplies of unprocessed wheat, corn, and beans.

For decades the Soviet Union has maintained its widely dispersed "iron rations" of wheat, which can be drawn upon only in time of war or famine. Soviet rulers will find much more credible a U.S. civil defense system that provides assured, readily available food stocks for evacuees.

The establishment of a strategic wheat reserve, stored where it would be readily available during and after a nuclear war, has been advocated over the years by several prominent Americans. President Kennedy recommended in 1961, without success, that some 126 million bushels of wheat be stored in 191 metropolitan areas. A bushel of wheat (60 pounds) will provide an austere but health maintaining diet for an adult for one month.

It would be a difficult and time consuming task during a crisis evacuation to distribute wheat stored in bulk in host areas to numerous shelters and buildings. The best proven way to store wheat in a readily transportable form is to seal the cleaned, dry wheat in a nitrogen atmosphere in a stout, plastic bucket.

Emergency baby food with long shelf life is an essential part of realistic preparations to maintain an evacuation or successful occupation of shelters. Most adults can remain healthy for several weeks without food if they have enough water; infants with nothing to eat can die in a few days.

The cost of a one-month supply of baby food, transported to and stored in host areas for one year, should come to no more than $18 per infant.

6. Blast shelters to protect on-shift policemen, firemen, and the reduced number of essential workers advocated herein should be mass produced and installed below ground in parks and other open spaces, well removed from buildings and from the worst dangers of fire and carbon monoxide to be expected in blast areas. (This was the fate of numerous Japanese trapped in shelters under Hiroshima buildings and the majority of the 135,000 Germans killed in the firebombing destruction of Dresden who died of carbon monoxide poisoning.) Using even strongly reinforced basements as shelters in areas subjected to nuclear blast and fire, as is currently planned by FEMA, would be disastrously hazardous.

   People are much more likely to commute to work in high risk areas during a crisis if they are provided with earth covered blast shelters. A corrugated steel blast shelter designed for mass production has been developed and successfully blast tested at 50 pounds per square inch in Defense Nuclear Agency's one-kiloton Misers Bluff non-nuclear test at White Sands, New Mexico. This shelter can be manufactured, delivered, installed, and equipped at a cost of less than $200 (in FY 1982 dollars) per person protected.

   Mass production models of austere concrete blast shelters, if likewise installed underground in open areas, may prove to be no more expensive. Development and blast testing should be expedited. If surveys show that 100,000 blast shelter spaces are needed for 200,000 persons working 12-hour shifts around the clock in high risk areas, the cost of the required blast shelters would be less than $200 million 1982 dollars.

7. Blast-tested designs of blast shelters should be made available to families and groups desirous of building private blast shelters. Concerned urban citizens are unlikely to give their strong support to a civil defense effort unless it provides them with a better hope for surviving a nuclear attack than they can anticipate if they evacuate to an unprepared countryside or to basement shelters away from their homes that would afford poor protection against radiation or blast and little or no protection against fire or carbon monoxide.
8. Detailed plans for the rapid construction of shelters and life support equipment during recognized crises of different durations should be developed for all communities. Such plans should include designs both of expedient shelters to be built during rapidly escalating crises and of permanent blast and fallout shelters for construction during a possible prolonged, recognized crisis. Local contractors should receive training and be listed together with local sources of earth-moving machines and shelter building materials. Athens, Tennessee, a town of about 35,000 people with very few buildings affording good fallout protection, has developed such a plan for building covered trench shelters for all its citizens in two days time.
9. Seven million additional sets of radiation monitoring instruments called for in FEMA's Program "D" should be produced

in the accelerated three-year program. Almost all of these essential instruments should be stockpiled outside probably targeted areas and preparations made for their distribution during a crisis.

Even if these radiation monitoring objectives are fully realized in three years, at least 20 million citizens who live too remotely from shelters to be officially designated for multifamily occupancy will have no instruments or other reliable ways to determine the changing radiation dose rates endangering them.

COSTS

The total cost of the civil defense preparations outlined above is estimated at about $4.5 billion. This total includes $2.6 billion (a FEMA estimate) for Program "D," about $1.64 billion for a one-month austere food supply stored in many parts of the host areas for 110 million adult evacuees and 8.8 million infants, approximately $200 million for blast shelters for persons who would work shifts in the mostly evacuated high risk areas, and about $30 million for the remaining changes and additions to Program "D."

For the first year after funding, approximately $300 million could be spent effectively; for the second year, $1.5 billion; for the third year, $2.7 billion.

For the modified three-year program, the annual cost per U.S. citizen would amount to less than $7, contrasting with a cost of about 50 cents per person today. Increasing U.S. per capita civil defense costs over tenfold would put U.S. costs into the range of what several Scandinavian countries are spending but still well below Swiss expenditures, which are about $32 per capita. The new U.S. civil defense program would represent per capita expenditures at the bottom range of the current Soviet costs, which currently run $8 to $20. The integrated Soviet civil defense costs over the years for construction, equipment, stockpiling, etc., run to thousands of dollars per Soviet citizen, in contrast to virtually nonexistent U.S. civil defense efforts.

CONCLUSIONS

The time is very late for initiating strategically significant American civil defense preparations. In future years, the difficulties are likely to increase. A start should be made soon, low key yet resolute, and build rapidly toward a credible lifesaving, nation preserving capability.

Once this realistic war-surviving project is underway, it will signal that the U.S. has abandoned the demoralizing policy of Mutual Assured Destruction and is determined to attain Assured Survival.

APPENDIX G: THE SOVIET VIEW*

In the Kremlin's perception, the "big question" about the High Frontier decision will not be whether the U.S. has the technological, industrial, organizational, or managerial resources and skills necessary to carry it through. The big question will be whether the U.S. leadership and the American people under that leadership have the will and resolve to do the things that are necessary to bring the nation's technological and industrial might to the service of this project.

Moscow, over the past decade, has increasingly convinced itself that the U.S. is in a state of steady decline, both domestically and internationally. A variety of factors has been adduced by the Soviets in explanation of this decline. They rest upon the panoply of stock-in-trade Marxist Leninist theories and concepts that are relative to the workings of immutable laws of social development. These laws dictate that the fall of the U.S.-led capitalist system and the triumph of the Soviet-led communist system on a world scale is historically inevitable. However, in terms of how the decline has proceeded, the Soviets have emphasized a succession of concrete instances in which the U.S. has exposed its lack of will to "act in old ways" and its failure to confront the USSR from "positions of strength." For example, the Soviets cite:

- The acceptance of defeat in Vietnam.
- The concession of strategic parity to the USSR, first in practice in the late 1960s and then formally in 1972.
- Acceptance of the Soviet-structured "principles of peaceful coexistence" in U.S.-Soviet relationships from 1972 onward.
- Repeated backdowns in the 1970s when U.S. and Soviet interests clashed at different points in the world.
- Consistent retrenchment in the purposeful development and use of the nation's scientific and technological potential to military ends.

By the end of 1980, the Kremlin appeared to genuinely believe: (1) that the U.S. had reached a point of no return in its decline, (2) that as a consequence, an irreversible shift in favor of the USSR had taken place in the "correlation of world forces" against the U.S., (3) that the U.S. could no longer pick and choose its policies and courses of action, and (4) that the U.S. has no choice but to recognize and act in accord with the "objective realities" of an adverse power situation in the world.

How does the Kremlin look at this U.S. situation and prospects after a year of President Reagan in the White House? Does it consider him as bound by immutable limitations imposed by an adverse balance of world forces? Or does it see him as defiant of "objective realities" as the Soviets define them and willing and able to follow courses as he sees necessary in the pursuit of the fundamental interests of the U.S.?

The answer here is central. It is crucial to the answer to the more specific question of how seriously the Kremlin will view a High Frontier decision by the President.

It should be noted that the Soviets have come to react with particular skepticism to declaratory or deliberately arranged revelations of new U.S. approaches in military affairs as a consequence of their new appraisal of the U.S. For example, a prominent, Soviet Washington watcher has said

* Unless otherwise indicated, all dates appearing in this Appendix are 1981.

that "it is very typical of American political tradition for new leaders to formulate some kind of new policy line, give it an eye-catching title, and thus 'declare' themselves, without, however, any readiness to do the things necessary to effect any meaningful change." In this context, Soviet military analysts lightly disregard such past stated U.S. doctrines as "massive retaliation," "flexible response," "graduated deterrence," "city avoidance," "the Nixon Doctrine," "sufficient deterrence," "flexible options," etc. They assert that all of these and others represent nothing more in real terms than empty subterfuges intended to suggest capabilities that do not exist. Directly on point, Soviet observers openly disdained announcements and contrived leaks during the last months of the Carter Administration regarding new weapons possibilities and new military strategies and programs.

The issue we now face is whether the Soviets continue to have such attitudes or whether after a year of experience with the Reagan Administration they now perceive Washington as sufficiently purposeful and resourceful enough to confront the USSR with a truly bold new departure in its efforts to refurbish U.S. military strength and to reestablish U.S. power and influence in the world.

At the present stage of the Reagan Administration, Moscow gives convincing evidence that it perceives a genuine turnabout underway in U.S. international purposes and attitudes and that it sees itself confronted by a formidable challenge of indefinite duration across the broad spectrum of global struggle. Further, it indicates that it is finding the present policies and courses of action by the new Administration to be decidedly disturbing from the standpoint of Soviet aims and expectations. Further, it sees itself in a different ballgame than it has engaged in with the U.S. over the past 20 years. It feels it is no longer testing the U.S. but is itself being tested by the United States.

The key point about present Soviet assessments is that the Reagan policies have profoundly shaken the solidly held Soviet conviction that, regardless of Reagan rhetoric before his election, the hard realities of the world power situation would force him to follow essentially the same policies toward the USSR as his three predecessors.

During his visit to Moscow in May 1972, President Nixon unexpectedly, from the Soviet point of view, accepted Soviet-defined principles of "peaceful coexistence" as the guiding rule for U.S.-USSR relationships and, at the same time, acknowledged and even welcomed Soviet attainment of strategic parity vis-a-vis the U.S. The Kremlin did not expect this and has since ridden a wave of increasingly high confidence.

In the Kremlin, perception of U.S. leadership made this profound shift not because of a sudden change of heart but because the U.S. was forced to do so by the "objective realities" brought about by a decisive shift in the "correlation of world forces" and primarily in the balance of military might in favor of the USSR. Subsequent developments in the U.S.-Soviet relationship brought a hardening of this view to the point of Soviet conviction.

By the time Reagan assumed the U.S. Presidency, the Kremlin clearly evidenced its settled assurance that it was fully in control of the continuing global struggle between the USSR and the U.S. In fact, it might be said that it viewed the future as its oyster. To be sure, the Kremlin figured that Reagan might prove troublesome in the beginning of his tenure, given his ideological credentials, but this would necessarily be short lived. Given the actual power balance in the world, like his immediate predecessors, Reagan would necessarily have to seek accommodation with the USSR.

Brezhnev had classically voiced such an assurance on the eve of the 1976 U.S. Presidential election when he told a Central Committee Plenum that "whoever might be in office in Washington . . . the United States will have to reckon with the actual alignment of forces in the world which in recent years have prompted the American ruling circles, after making a sober analysis of the existing situation, to seek ways of coming to an understanding with the world of socialism."

Experiences with Carter were obviously seen by Moscow as fully confirming Brezhnev's predictions. After Reagan's 1980 victory over Carter, Soviet authorities hastened to apply the same thinking to the new U.S. leader. On the day following the election, Alexandr Bovin, a leading Soviet commentator on international affairs, contended that "any U.S. president is guided not by any personal ideas and ambitions, but by the realities of the world today . . . This is undoubtedly true, too, of the future government of President Reagan."

When the early policies of the Reagan Administration appeared to run counter to what the Soviets considered to be the dictates of the balance of world forces, Soviet spokesmen insisted that it would be only a matter of time before the necessary corrections would be made. Thus, *Pravda* asserted on March 9, 1981, that time was running ning out for the Reagan Administration: "It will soon have to make its position clear on a number of very important political issues . . . The situation since the 26th CPSU Congress has merely laid bare still further . . . the yawning chasm in its [the U.S.'s] basic policies and the objective realities of the modern world situation." And on March 22 *Pravda* spoke even more directly: "The captains of Western policy will sooner or later have to heed the inexorable realities of our time . . . No one is going to turn back the wheel of history."

However, as Washington persisted in policies not to Soviet liking, Moscow appeared increasingly to accept the fact that the new U.S. Administration was, in fact, demonstrating an unexpected imperiousness to limits on its freedom of choice that were supposedly imposed by the "objective realities" of an adverse "correlation of world forces." Leonid Zamiatin, head of the CPSU Central Committee International Information Department, told a Radio Moscow audience on May 16:

> Yes, comrades, we are now again running into facts which indicate that among those who determine the policy of capitalist countries are many officials who would like to test our durability, who are not leading affairs toward a qualitatively new situation in the world—toward creating a situation of confrontation . . . I have in mind the new Reagan Administration, which recently came to power in Washington. The new administration considers that opposition to the Soviet Union . . . in the economic, political, and other fields is its main policy concept. Besides, they maintain that this opposition must be on a global scale.

Soviet Politburo member and Defense Minister Dmitri Ustinov gave confirmation of such an appraisal at the highest level of Soviet authority in a truly remarkable article that was given striking prominence in *Pravda,* the main organ of the Soviet Communist Party, on July 25.

Ustinov observed with obvious satisfaction that "it proved possible in the 1970s to channel international relations into the relaxation of tension and to commence restructuring them on the principles of peaceful coexistence." He detailed the benefits that had derived from the Soviet standpoint as a result; but none of this, he then lamented, holds good with the Reagan team:

> At the turning point of the 80s there has been a radical change in the policy of the United States and a number of other NATO countries. The upper hand there was resumed by circles which orient themselves on force in international matters, which refuse to accept the changes conditioned by history which are taking place in the world, and which have set

for themselves the objective of altering, in their favor and at any price, the balance of power in the world arena.

He followed by bearing down on specific transgressions of the new Administration in Washington and then suggested bafflement—indeed disbelief—at what the Kremlin purports to see as a foolhardy defiance by the U.S. of the "hard realities" it faces with the present adverse balance of power. He asked, "Where is such a course leading? What are the new U.S. leaders preparing for the world? The ruling circles of the United States are bringing matters to a further intensification of international tension and an exacerbation of Soviet-American relations." In a further question, he suggested very real apprehension on the part of the Kremlin, albeit under cover of standardized Soviet bravado. "Is the White House really thinking seriously of reviving the 'cudgel policy' in the 1980s? The world today is not as it was. It is not they alone who have force at their disposal."

According to its own testimony, what is most disturbing to the Kremlin is that the U.S. Administration is evolving a comprehensive international policy and program of action which treats the Soviet threat as central and, on the other hand, avoids direct engagement of the U.S. with the Soviet Union.

For the past 20 years, Moscow has been accustomed to U.S. administrations being preoccupied with *relations* with the USSR. This is the first time since the 1950s that it has had to deal with one preoccupied with *opposition* to the USSR. In this connection, Soviet spokesmen make clear that the Reagan postures and policies are having telling effects upon the Kremlin's feeling of assurance that it is in the driver's seat and knows where it is going and by what means. Rather, the Reagan postures and policies are producing signs of uncertainty, if not confusion, and a seeming loss of a sense of direction in the Kremlin. At the same time, these spokesmen leave no doubt that the Kremlin is assessing the new U.S. Administration as fully resolute in its purposes and knowing in its plans and, hence, both intent upon and capable of major efforts to bring about a reversal of roles in U.S.-Soviet relationships.

A number of factors and circumstances are evidently responsible for this situation:

1. *The Administration's Pervasive Focus on the Soviet Threat.* Soviet spokesmen at all levels are constantly talking of present U.S. attitudes and policies as being taken up with "antiSovietism," "anticommunism," and "the Soviet threat." The Soviets themselves have, of course, always stressed, indeed made a virtue of, their own focus on the antiimperialist (i.e., anti-U.S.) struggle in virtually everything they do and plan to do. They do not like the shoe on the other foot. *Izvestiia* reflected the Soviet concern on May 21 when it spoke of "the anti-Soviet hysteria which the Reagan Administration is stoking." As noted above, Zamiatin contended on May 16 that "the new administration considers that opposition to the Soviet Union . . . is its main policy concept." Marshal Kulikov, commander of Warsaw Pact forces, wrote in *Red Star* on June 21 that U.S. concentration on hostility toward the Soviet Union is similar to "the aims set by the Hitlerite leadership in attacking the Soviet Union," and that anti-Soviet statements "are not dropped by chance. They express the definite political thrust and activity of the new U.S. administration." Marshal Ustinov spoke in his *Pravda* article of July 25 of "unconcealed anti-Sovietism" on the part of the Administration, and added that, "The U.S. has chosen anti-Sovietism at the top of the list of things that currently form 'the basis of its foreign policy.'"

   It is in such a light that the Soviets view President Reagan's decision to reverse President Carter's abandonment of the linkage concept. Soviet spokesmen interpret this move as demonstrative of an intent to force fundamental changes in Soviet policies and conduct. Politburo member Konstantin Chernenko singled this out for special attention in his Lenin Day speech on April 22: "The United States stubbornly asserts that talks on certain specific questions must be linked with the whole gambit of international problems." Foreign Minister Gromyko dwelt at length upon the alleged evils of the concept in a major article he wrote for *Kommunist* in January, saying, among other things, that "with this linkage it is essentially impossible to resolve a single international problem."

Another specific of the Reagan Administration's anti-Soviet posture that is perturbing to Moscow is that, unlike the situation with the Carter Administration, the President and his advisors have shown they do not intend to be influenced in their choice of policies and courses of action by concerns over displeasing or provoking Moscow. Outstanding cases in point from the standpoint of impact on the Soviets have been the Administration's stand on the Cuban-Soviet involvement in El Salvador, the Administration's expression of open opposition to the Soviet threat to Poland, and the Administration's disregard of Soviet sensitivities over its moves toward closer relations with China, moves which *Pravda* openly viewed "with alarm" and characterized as confronting the USSR with "a new and highly dangerous stage in the development of the Chinese-American partnership."

Finally is the matter of calling a spade a spade. Something of a shock wave has been produced on the Soviets by the practice of the new U.S. Administration of speaking frankly about the motives, purposes, policies, behavior, and prospects of the Soviet leadership. When President Reagan spoke at his January 29 press conference of Kremlin aims to establish a world communist order and added, for good measure, reference to the peculiar Soviet rules of morality, screams of outrage were sounded by Moscow. So, too, when Secretary of State Haig charged the Soviets with using international terrorism in their conduct of foreign policy. Similar reaction met President Reagan's assertion in his May 17 address at Notre Dame that the West will simply transcend communism, that "it will dismiss it as some bizarre chapter in human history whose last pages are even now being written." While some Americans, and other Westerners, have deplored such frank talk by the President and his team, the Soviets in their reactions have made clear they perceive at risk here a one way advantage they have long enjoyed in the conduct of ideological warfare against the U.S. —more than this, by a foe no longer willing to tie its hands when engaged in a fateful struggle for its survival. That, the Soviets make clearly evident, troubles Moscow no end.

2. *Downplay of Dialogue and Negotiations.* The Soviet perception that the Reagan Administration is proceeding with its international programs without directly engaging the U.S. with the USSR has been most starkly confirmed by Washington's response, or better by its lack of response, to mammoth Soviet efforts to get it involved in various negotiation tracks. Moscow makes it crystal clear that it is deeply troubled about the consequences and implications.

Moscow, by way of Brezhnev personally, has served up a heavy menu of proposed talks, agreements, and negotiations to the Reagan Administration. Initially, the Kremlin's purpose was clearly to involve the U.S. once again in the same sort of negotiating gamesmanship that had proved so useful as a means of forestalling U.S. initiatives with previous U.S. administrations. None of the proposals was new or involved any sort of change in long held positions on Moscow's part. All were designed to put responsibility for concrete responses on

Washington to force the U.S. either to get involved in exchanges and negotiations with the USSR for problem areas and on terms of the Soviets' choosing or, if it did not respond or responded negatively, to expose U.S. recalcitrance and aggressiveness in contrast to Soviet peace loving policies and intentions.

How it was all supposed to work has been well explained by the Soviets themselves. Thus, Candidate Politburo member and Central Committee Secretary Boris Ponomarev wrote in *Kommunist* in March: "The USSR's peace proposals raise a dilemma for American ruling circles: either to positively approach the peace proposals or to demonstrate their disregard for people's aspirations, and thus be viewed as warmongers."

However, as time passed and the Reagan Administration proceeded with development and implementation of various elements of its international program without talking matters over with Moscow or responding in any other way to Soviet efforts to involve the U.S. in dialogues and various negotiating tracks, the Kremlin began to indicate deep concern.

Pravda complained on March 24 that the "new Government appears to be bent on . . . not a lessening of international tension but its growth." On May 4, the party newspaper added that "there have been astonishingly few new actions or proposals" of the Reagan team relative to the USSR, and on May 31 it went further and declared "in slightly over four months the Republican administration has not advanced a single position proposal on a single important problem." Meanwhile, Brezhnev added his voice to the complaints. At Kiev on May 9, he asserted that "the solution of international problems by way of talks and mutually advantageous agreements appear to be way down on their list of priorities, if they give serious throught to this at all." And in a Kremlin speech on June 10, he said:

> I can say quite definitely: Not a single real step has been made by the United States so far during all the time since the advent to power by the present U.S. administration either on that [Euromissiles] or other questions in order to continue, at least in a preliminary order, the discussion of the essence of these questions. On the contrary, the Americans are delaying on various pretexts the beginning of such a discussion while we, for our part, are prepared for it at any time.

3. *A "global U.S. offensive" to reestablish positions of strength worldwide.* Throughout Soviet assessments of the U.S. under Reagan there is an oft-evident Kremlin perception that (1) in lieu of seeking agreements with the USSR, the new U.S. Administration is following its own agenda to develop positions of strength wherever U.S. and USSR interests clash and (2) it is achieving some significant successes in this regard.

Pravda reflected such a perception as early as March 25, asserting that "the White House can find nothing better than to settle the most complicated and acute problems of international life by means of arms and positions of strength." Similarly, *International Affairs,* the leading Soviet foreign affairs journal, claimed in April that:

> Reagan and other highly placed Washington officials in essence have thrown down a challenge to the whole world community . . . In a word, the bosses of the White House by their adventurist actions are trying to introduce chaos and confusion into international relations to the point of exacerbating the situation to unravel the legal-treaty system and to put the world on the brink of a nuclear-

missile catastrophe.

In developing its analysis, this *International Affairs* article clarified that what is disturbing Moscow is that it sees the Washington government systematically undertaking to redress the shift in the "correlation of world forces" which the Soviets have viewed for several years as having moved irrevocably in favor of the USSR.

The second ranking Soviet military leader after Defense Minister Ustinov, Marshal Nikolai Ogarkov, chief of the Soviet General Staff and Deputy Minister of Defense, in a highly significant article in the July issue of *Kommunist* charged that U.S. international moves "are linked by a single design and have as their goal the gradual consistent weakening and undermining of socialism as a [global] system and, as a result, the establishment of the world rule of U.S. imperialism."

4. *The Reagan commitment to redress the military balance.* Underlying and giving indispensable, substantive weight to all other Soviet concerns about Reagan policies is the Kremlin perception that the President intends an unrelenting effort to refurbish U.S. military power as an indispensable prerequisite for an effective policy toward the USSR. This portends for the USSR not just a new arms race challenge but, far more important, a new challenge to the total of Soviet global objectives and expectations.

At the 26th Party Congress, Brezhnev on February 23 took special note of President Reagan's military plans, asserting that "military expenditure is growing at unprecedented rates" in the U.S. and NATO and that "a considerable part of these vast resources is being spent on the accelerated development of new types of strategic nuclear weapons." The Soviet leader has repeated the same theme over and over in later public pronouncements. Thus, typically, he said in a major speech in Kiev on May 9:

> But there are also such statesmen in the bourgeois world who, judging by everything, are accustomed to thinking only in terms of strength and diktat. They actually regard the attainment of military superiority over the Soviet Union as their main political credo.

Other top level Soviet authorities, particularly military, have not only echoed Brezhnev charges about the present and intended scale of the new U.S. military efforts, they have added strong notes of alarm about the warlike purposes of the effort.

Thus, Defense Minister Ustinov writing in *Pravda* on February 21 argued that the U.S. is not only engaged in an all-out military buildup but is preparing to use its new programs for offensive actions against the USSR and its Warsaw Pact allies. In Ustinov's words: "Imperialist propaganda speculates on the false thesis that the USSR is allegedly in an excessive buildup of nuclear missile forces. But this buildup is in fact taking place in the United States, where more and more new generations of ground and sea based nuclear missiles are being developed and produced and plans have recently reemerged for the production of neutron weapons and their deployment in Europe. The Pentagon is counting on nuclear weapons for attaining U.S. global strategic goals by delivering preemptive nuclear missile strikes against the Warsaw Pact countries."

Marshal Ogarkov, Chief of General Staff, wrote in *Red Star* on May 9 that the United States "is striving . . . to clear the way for an uncontrolled arms race . . . with the aim of direct military preparations." He concluded that "one cannot fail to notice their definite resemblance to events of the 30s."

Marshal Ustinov in his July 25 *Pravda* article put the cap on all such statements of concern: "The present administration has stated that the attainment of U.S. military superiority over the Soviet Union is its foremost aim in the next few years. To this end, military spending next year by the Department of Defense alone will increase by more than $40 billion and will reach $226 billion. Altogether during the next five years (to 1986), the United States intends to spend $1.5 trillion, that is almost as much as it has spent on the armed services in the last 12 years. The question arises: Why does the United States need such enormous military expenditure?"

Ustinov neatly answers his own question and in doing so starkly reveals one of the major worries of the Kremlin: that the U.S., through its new military buildup, will break out of deterrence constraints and position itself to use force to protect its interests and positions at any point where they may be threatened in the world.

> This whole course of the White House, which is dangerous for the cause of peace, is aimed at giving itself an excuse to react to possible conflicts occurring in any part of the world by means of military force. Not since the days of the cold war has the line of using force shown itself so plainly in U.S. policy. 'We must restore the high mobility of the armed services,' Secretary of Defense C. Weinberger asserts, 'and react quickly to changing situations in any part of the world—we must strengthen our positions in the world by means of arms. . .'

In the face of its reading along the above discussed lines of the policies and purposes of the U.S. under its new President, the Kremlin appears to be readying itself for a trying struggle with the U.S. both for the short and long haul. It gives evidence that it no longer expects, as it did so strongly before Reagan entered office, to be able to ride the momentum of past successes to new and ever more decisive gains over a foe in retreat and with little evidence of confidence in its ability to end that retreat.

Still, however, Moscow does not suggest that it sees reason to batten down and wait for more propitious times. It professes to believe that the USSR continues to enjoy important advantages over the U.S., that these advantages cannot only be preserved but extended, and that thus the USSR can aggressively continue advancing across a global spectrum. As an overall matter, it insists that the "correlation of world forces" is now "definitely and irreversibly" in its favor and avows an ability to bring about further shifts in its favor.

As a preferred way of proceeding in the existing situation, Moscow makes it clear that it is still ardently seeking, and will unquestionably continue to seek, the revival of a Soviet-style detente relationship with the U.S. As a result, it would be able to press forward toward its own goals while the U.S. subjects itself to self-imposed restraints in the interest of an illusory, indeed hopeless, stability.

Brezhnev on June 9 set forth the deal, with all the usual blandishments, which the Kremlin is still trying to sell the Reagan Administration and which it will doubtless try time and time again to sell it in the future.

> For its part, the Soviet Union put forward at the recent CPSU Congress a broad program of specific measures to create a healthier international climate and to build up confidence between states. We will pursue this policy persistently and with consistency. Of course, this also concerns our relations with the United States. We tell the U.S. leaders in our contacts with them, and I am repeating it in public: We do not seek a confrontation with the

United States, we do not encroach upon America's legitimate interests. We want peace, cooperation, and normal relations between our countries based on mutual trust. It is precisely why we offer the United States and the other Western countries fair, constructive talks, a quest for mutually acceptable solutions to practically all the major issues existing between us. We are for a joint quest for ways toward a lasting peace and mutually acceptable solutions to practically all the major issues existing between us. We are for joint quest for ways toward a lasting peace and mutually beneficial cooperation.

But the Kremlin is serving notice in many different ways that unless and until this deal is bought, which it appears now to consider highly unlikely, the USSR will proceed with its own program of action to combat the U.S. at any and all points of contact and with what it considers tried and true methods and means.

THE SOVIET SITUATION WITH RESPECT TO A NEW ORDER OF ARMS RACE WITH THE U.S.

Moscow makes it clear that what happens in the military field is central to everything else. As it so often says, it sees the growing military might of the Soviet Union as the indispensable underpinning of the favorable correlation of world forces.

Moscow categorically insists that the USSR can and will do everything necessary to prevent an alteration in the present relative military positions of the two countries. It professes confidence that it can succeed in its efforts.

The Soviet posture today on the military balance issue remains as it was defined by a resolution adopted by a Central Committee plenum in June 1980:

> Detente is the natural result of the correlation of forces in the world arena that has formed in recent decades. The military-strategic balance between the world of socialism and the world of capitalism is an achievement of principled historic significance. It is a factor containing imperialism's aggressive aspirations which meets the vital interest of all peoples. Hopes of shaking this balance are futile.

To a certain extent the Soviet leadership is drawing assurance from the time factor. Thus it was recently said: "No matter how hard the American civilian and military leaders try to frighten the Soviet Union with their 'well-orchestrated' strategy of counterforce superiority, the United States does not possess this kind of superiority because the majority of the systems on which (the strategy) depends will not be ready for use until the second half of the 1980s."

Beyond this, the top Soviet political-military leaders in foreign policy-security areas insist the USSR is determined to keep moving in every branch of military activity.

At the Party Congress in February, Brezhnev repeated the standard Soviet line that the USSR does not seek superiority, but "neither will we allow such a superiority to be established over us." Speaking in Kiev on May 9, Brezhnev said "if we are compelled, we will find a quick and effective response to any challenge by belligerent imperialism." And specifically about the European theater, which presently stands at the top of Moscow's agenda, he said on May 22 in Tbilisi:

> I must say with a full sense of responsibility that we cannot leave without consequence the deployment on European soil of new American nuclear missiles aimed at the USSR and our allies. In this case we will have to think about extra defense measures. If necessary, we shall find impressive means to safequard our vital interests. And then the

NATO planners should not complain.

In their repeated assertions that the USSR will keep pace in any arms contest with the U.S., no matter what its nature, scope, or duration, Soviet authorities categorically insist that the Soviet economy can stand up to any requirements that may be placed upon it. Indeed it can outlast the U.S. economy in a no-holds-barred arms race. Typical of a vast number of pronouncements along this line is a May 1981 article in a leading Soviet political and economic journal:

> It goes without saying that the Soviet Union and the other fraternal countries must take all the necessary measures to reliably defend their socialist gains, which involve diverting resources from creative goals. It is well known, however, that in the last decade the CEMA members' economic growth rates were double those of the developed capitalist countries. The Soviet Union and the socialist community are capable of accepting and enduring any competition with imperialism in the military-economic sphere, although, of course, they would prefer, as would the peoples of the whole world, to expend their resource solely on creative aims.

Despite such stated assurances, however, current performance of the Soviet economy and a recent analysis of the economic impact of excessive Soviet military expenditures indicate that realistic possibilities may be quite different.

The Central Statistical Adminstrations of the USSR released data on July 25 on results of the first six months of the Eleventh Five-Year Plan. It showed the USSR as increasingly beset by serious economic strains.

This data followed an earlier published Ministry of Defense monograph by military economist A.I. Pozharov, *The Economic Bases of the Defense Might of a Socialist State,* which considers the question of the limits beyond which the Soviet economy cannot go to serve military purposes among a variety of other things.

Together, the monograph and the statistical data raise serious doubt about the official optimism of Soviet leaders on Soviet capability to meet any new challenge the U.S. may offer in the arms competition field.

The general line of these Soviet leaders regarding arms competition is that the high level of Soviet economic and technical resources, plus advantages of the "socialist system," ensure Soviet ability to match the West at any level of strategic and conventional arms competition. This is asserted despite the convincing evidence that the Soviet Union is already spending a substantially larger proportion of its GNP on defense than the United States and that Soviet defense spending is growing at a rate higher than Soviet national income.

The Pozharov book and actual Soviet economic performance, however, indicate a radically different situation than that depicted by Soviet leaders. Pozharov indicates that there are limits to Soviet defense expenditures which, if exceeded, would result in serious damage to the general economy upon which the defense capability of the USSR directly depends. Other works by military economists imply that these limits have already been reached. Pozharov also cautions that excessive military manpower draws off productive workers from both the general economy and the defense industries, thereby hindering the productive capacity of both.

As for actual Soviet economic performance, Soviet statistics show: (1) that the recent Soviet Five-Year Plan for 1981-1986 has gotten off to the worst start during its first six months of any five-year plan since the war, (2) that this follows performance under the previous five-year plan that was more disappointing than any plan in Soviet history, and (3) that it comes after some two decades of a steady decline in the rate of growth of the Soviet economy, which appears likely to fall to less than three percent per year soon.

For the past several decades, the Soviet leadership has been able to maintain a steady growth in military expenditures as a result of the relatively high, albeit declining, growth rate of the economy as a whole. Since the mid-1960s, it has been aided in this by the failure of the USSR's opponents, chiefly the U.S., to present it with special challenges in either the armaments or political field and has been, in consequence, able to set its own priorities in accord with its own determination and ordering of its objectives.

Now, however, the policies of the Reagan Administration, particularly with respect to the U.S. military buildup, confront the Kremlin with a serious problem of resource allocation.

The issue raised by this resource allocation problem is not just the standard conflict between military requirements and consumer requirements. The rate of growth for consumer goods under the current five-year plan is, as always, extremely modest, and were it completely wiped out, or even substantially reversed, the impact on the overall economic situation of the USSR would be minimal at best.

The issue is whether an increased proportion of the resources is to go into the production needed for ensuring maximal battle readiness of Soviet forces in the face of growing U.S. preparedness as against the proportion of resources going into the continuing development of the sinews of Soviet power, heavy industry, and defense industries.

Resource allocation problems traditionally have been the source of intense policy disputes within the Soviet leaderships. The problem that appears looming now can surely be no exception. Indeed, there is good reason to assume that in the documentary materials being emitted, such disputes are already underway.

In any event, it is clear that indicators along this line, as well as other indicators relative to the Soviet economic situation, will require close watch to determine the extent to which U.S. military expenditures prove a source of mounting pressure on the Soviet leadership.

The Pozharov monograph, which is billed as "intended for officers, generals, and other studies of military economics," warns that excessive military allocations will indeed undermine the foundations of Soviet military power. The author argues that excessive military expenditures "could decelerate the development of the very bases of military power—the (general Soviet) economy—and therewith inflict irreparable damage on the defense capability." Thus, this passage as a whole says:

> Military-economic power for a socialist state is not an end in itself, it does not automatically follow the growth of economic power, but is reflected in the real needs of society for military power. A sharpening of the international situation forces a socialist state to increase military production and consumption, relaxation of tension permits their reduction, more fully to utilize economic power for raising the well-being of the workers and development of the economy. However, it is impossible to permit, on the one hand, a lowering of military-economic power because in this case it would threaten the defense capability of the country, on the other—too much of an increase because in the final analysis this can decelerate the development of the very basis of military power—the economy—and inflict thereby irreparable harm to defense capability. 'If a war is lost because of overstraining the economic power of a country'—wrote B.M. Shaposhnikov—'then such a loss can be received even before the beginning of the war as a result of a high military budget, whose burden does not correspond to the capacity of the population to pay and when the military budget does not go hand in hand with the economic development of the state'.

The author likewise contends that excessive military manpower has two adverse effects. First, it decreases the pool of workers available to the civilian sector, thereby limiting the potential for general economic development. Second, and partly as a result of the first, it decreases the workforce and production equipment available to meet warfighting production requirements.

In sum, Pozharov takes issue with what he calls "the popular expression that three things are necessary for war: money, money, and even more money."

The growing decline in the growth rate of the Soviet economy over the past two decades (that is, over the period of the sustained all-out concentration of the Soviet leadership on the buildup of Soviet strategic power) would appear to indicate that military expenditures are already approaching, if not exceeding, "the objective limits" beyond which the USSR cannot go without serious damage to the economy as a whole, including the reproductive base crucial to the very existence of Soviet military might.

Soviet sources make clear that the poor prospects for Soviet industrial growth can be attributed to the same complex of factors that have plagued the Soviet economy for a succession of five-year plans; growing constraints on manpower, capital, and energy; built-in obstacles to technological innovations; rising difficulties in acquiring and distributing raw materials; inflexible and inefficient planning and management system; low labor productivity; and, of course, the rising burden of defense.

There is another aspect of a thoroughgoing U.S. arms race challenge to the USSR that raises a serious question regarding official Soviet optimism: that is the possibility that the U.S. will make full use of its industrial and technological superiority.

It is also important to stress that despite a deliberate effort by the Soviets to downplay the possibility that any meaningful technological breakthrough in weaponology can or will be effected, both political and military authorities give indisputable evidence that Moscow is, in fact, deeply fearful of just such an eventuality. A warning sounded back in 1970 by Brezhnev at the conclusion of the Dvina military exercises provides a dependable base point for Soviet thinking:

> We now have fine equipment, but as is known, we live in an age of scientific and technical progress when weapons are being improved so rapidly that new forms and systems are often created not just within a year but within a shorter period. Our scientists, both civilian and military, must constantly think about this and remember it.

In later years, Brezhnev's warning has been repeated or embellished many times. In 1976, for example, a major article in *International Affairs* argued that military hardware "must either keep pace with the rapid development of the scientific and technical revolution, constantly absorbing its latest achievements in every field . . . or risk being converted into a pile of rubbish."

And a foremost Soviet student of the impact of science and technology on military affairs wrote in *Communist of the Armed Forces* in September 1974:

> In as much as there are no limits to understanding natural laws, so there can be no limit to the application of these laws in technological designs. From this point of view, any, the most terrible, weapon cannot be called absolute since in its stead can come a still more powerful one based on the newest scientific-technological achievements.

The same view was expressed earlier as sort of basic doctrine in an editorial article in *Communist of the Armed Forces* back in 1966:

> Achievement of military-technological superiority by one side or another does not guarantee continuation of this superiority in the future. The stern dialectic of development consists in the fact that the struggle for superiority must be waged constantly. Any weakening of effort in this field, any excessive self-admiration because of the successes achieved, can lead to the loss of this

superiority.

Brezhnev at the 26th Soviet Party Congress directly emphasized Soviet concern over possible U.S. weapon breakthroughs by the U.S. in its current military effort when he proposed "limiting development" of Ohio-class submarines and banning new weapons for such submarines. This particular proposal represented the resurfacing of one element in a broad ranging proposal for a "ban on manufacturing new categories of mass destruction weapons and new systems of weapons" which Brezhnev first asserted in June 1975 and which has subsequently been the object of extensive Soviet diplomatic and propaganda activity.

Specifics of the proposal as well as the appreciable number of subproposals the Soviets have derived from it have all been directed toward prospective U.S. weapons developments. Initially the focus was on Trident submarines, the B-1, and cruise missiles. It is evident that the focus is now shifting to prospective advances by the U.S. in the military use of space, in ABM technology, and in the advanced weapons needed to implement such new U.S. doctrines as those set forth in Presidential Directive 59.

Relative to a new order of technological challenge to the USSR, these latter considerations almost certainly temper the Kremlin assessments of the problems they may ultimately face as a result of the U.S. President's arms program. They make it highly likely that the Kremlin views the situation as involving something far more than that of outstripping or keeping up with the U.S. by doing more of what each is now doing. It would seem in particular that they must raise in Kremlin eyes some alarming possibilities for the USSR, such as:

- Being forced into lines of effort the Kremlin did not choose as a result of new types of U.S. military capabilities or, at minimum, having to change its elaborate and complex war-readiness and war-fighting plans.
- Having its industrial and technological resources taxed beyond their capacity.
- Seeing fractured its basic strategic design of keeping the U.S. so boxed in militarily as to make it irrational for it to use its nuclear power for anything other than a self-defeating retaliation against a massive blow on its own territory.

THE SPECIFIC QUESTION OF THE IMPACT OF HIGH FRONTIER ON THE USSR

Viewing Kremlin apprehensions over the possibility of a radical new departure on the part of the Reagan Administration in the arms field, the Soviet reaction to a decision for High Frontier must be assessed.

Moscow is already indicating grave concern that the U.S. may be preparing to use space in its efforts to reverse the present imbalance in military power.

The first successful flights of the U.S. Space Shuttle in April and November of 1981 have served as the points of departure for literally hundreds of authoritative Soviet commentaries on alleged U.S. preparations, as *Izvestiia* typically put it, "to create and deploy in near Earth orbits, new generations of space weaponry systems . . . designed for carrying out strikes against targets in space, the atmosphere, and on Earth . . . in the hope that the U.S. will be able to avoid nuclear retaliation."

In the Soviets' view, High Frontier would confirm its worst fear about U.S. military purposes in space. It would view the move as what the Soviets themselves have characterized as a possible "absolute weapon" capable of ensuring U.S. "invincibility" from missile attacks. While the Kremlin would naturally consider fulfillment of this aim as some years away, at best it would, knowing the state of the technological art involved and of U.S. capability to build upon existing technology, still allow that substantial capabilities would be in place within a relatively short term of two to four years.

The Kremlin has always looked at space in terms of its military utility. Even as the USSR gained and exploited its unexpected dividends from the worldwide psychological impact of the first Sputniks, Soviet work programs were concentrated not on the Sputnik phenomenon, as such, but on using space to give added and perhaps decisive substance to the nuclear rocketry strategy that was then emerging from the "revolution in military affairs" that was said to have been effected in the USSR.

Khruschev spoke with surprising candor to President Kennedy about the intimate tie-in between space and military activities in the USSR at their meeting in Vienna in June 1961. He told the U.S. President that space cooperation was "impossible until there is disarmament" because of that tie-in. He said that there had been few "practical uses of outer space launchings," that these were "primarily for prestige purposes," and that such endeavors as an attempted "flight to the Moon" might weaken Soviet "defense" efforts.

The Soviet focus was already overwhelmingly centered on the near Earth environment, with the ultimate objective a system of multimanned, multipurpose space stations. And what was begun under Khruschev has been continued and greatly expanded under Brezhnev. Brezhnev explained the settled Soviet approach on October 22, 1969, saying: "Our country has at its disposal an extensive space program calculated for many years to come. . . . We are going our own way, and we are proceeding consistently and purposefully." Brezhnev then spoke glowingly of prospects of an entirely new Soviet reach in space: "orbiting stations will provide a highway into outer space—they can become 'cosmodromes in cosmos,' launching ramps for flights to other planets." Although left unmentioned by Brezhnev but frequently made explicit by other Soviet writers, the "cosmodromes" would give the USSR the means to command the near Earth environment for military purposes, with inestimable consequences for the strategic balance. And the USSR has over the last decade diligently and effectively pursued the goals set by Brezhnev. Indeed, it now appears close to the point where it can hope to serve up a possibly shocking space surprise for the U.S., and the entire Western world, if things continue as they are.

Meanwhile, the Soviets have consistently charged that U.S. space activities are directed toward military ends. This has been good propaganda from Moscow's standpoint, serving not only to further its efforts to stamp the U.S. with an indelible threat-to-peace image but also to blunt the impact of U.S. successes in space and provide a counter to attention to the military character of the Soviet program. But more has been reflected in such charges than propaganda. Implicitly reflected has been a genuine fear that the U.S. might in fact be using its technological prowess to beat the USSR in using space for military purposes.

Such a fear is certainly manifest in Soviet agitation over the military potential of current U.S. activities and plans for space. Early this year, in a journal intended for strictly internal Soviet consumption, it was asserted that "the Pentagon sees an important means for attaining military superiority over the Soviet Union and the other socialist countries in the creation of a powerful grouping of military resources in space." Unlike earlier periods when allegations about U.S. military aims in space were in very general terms, Soviet spokesmen now tend to be very specific in detailing actual U.S. plans, programs, and purposes.

It is in this light of Soviet attention to such specifics that the Soviet proposal which Gromyko submitted to the United Nations in August 1981 for a treaty banning the deployment of *any* weapons in outer space must be assessed. Moscow now claims that it has favored an agreement of this type since 1958. As a matter of fact, it rejected a proposal for such an agreement put forward by President Eisenhower in a letter to then Soviet

Premier Bulganin on January 12, 1958. In a speech prior to Moscow's reply, Khruschev disdained the U.S. proposal, saying: "This means they want to prohibit that which they do not possess." The formal reply in February 1958 conditioned Soviet acceptance of the proposal on Western agreement "on the prohibition of nuclear and hydrogen weapons, the cessation of tests of such weapons, and the liquidation of foreign military bases in the territories of other states." On March 15, 1958 Moscow submitted a proposal calling for "a ban on the use of outer space for military purposes" to the General Assembly of the United Nations. This ban was tied to the simultaneous "elimination of foreign military bases on the territories of other states, primarily in Europe, the Near and Middle East, and North Africa."

In all subsequent U.S.-Soviet negotiations and United Nations discussions of propositions relative to a total ban on the military use of space, the USSR stuck to its position that agreement on such a ban could be reached only in the context of U.S. acceptance of Soviet disarmament proposals. After years of wrangling, in 1963 Moscow did agree to join with the U.S. in a United Nations resolution calling for a treaty banning the deployment of weapons of *mass destruction* in space. Even then, four years were required before a treaty to this effect (the Outer Space Treaty) could be concluded and ratified.

Only now has the Soviet Union come up with a proposition for a total ban on weapons in outer space. The link between this break with the past and the Kremlin fears regarding new U.S. use of space to redress the existing strategic balance was explicitly spelled out by the Soviet delegate to the United Nations on occasion of the official tabling of the Soviet proposal at the First Committee of the General Assembly on October 21. According to a Tass dispatch of that date:

> Explaining the need for the proposed measure, Soviet delegate Vladimir Petrovskiy pointed out that the danger of the spread of the arms race to outer space was growing dramatically. Judging by press reports, the Pentagon is drifting plans of combat operations for control of outer space and planning to develop antisatellite weapons systems, including the establishment in outer space of military bases armed by such systems and the development of antisatellite mines. Work has been going on for some years now to develop laser weapons using rays of particles that would be based in outer space and on the Earth. The programs for launching manned reusable spaceships also include military aspects. The Soviet delegate said that the U.S. attempts to turn outer space into another scene for the arms race are explained by Washington's striving to secure superiority in armed force and intended primarily to establish its hegemony on the Earth.

Certainly one aspect of this Soviet move is strictly propagandistic. This should not be allowed to obscure the fact that the Kremlin is manifesting deep concern to avoid a weapons contest with the U.S. that would center on the use of high technology capabilities to establish an effective military presence in space. The last thing the Kremlin wants is a trial of strength in this area. Indeed, it can be safely said that to be in the area of high technology required for advances in weaponology, the Kremlin perceives their greatest single vulnerability as against the U.S. There can be no doubt that the Kremlin would view High Frontier as a challenge involving just such a trial.

The Soviets in countless ways have made it clear that they are keenly aware of the U.S. capabilities to accomplish even the most difficult tasks when it seriously sets itself to those tasks. In 1924, Stalin voiced what has been repeatedly demonstrated to be a fundamental Soviet appreciation of the U.S. in this regard. Stalin wrote, "American efficiency is that indomitable force which neither knows nor recognizes obstacles; which with its business-like perseverance brushes aside all obstacles; which continues at a task once started until it is finished, even if it is a minor task; and without which serious construction work is inconceivable."

Khruschev spoke to Kennedy at Vienna in much the same terms. And the Soviets continue to openly acknowledge that the U.S. "occupies the first place in the world in terms of the general development of science and technology, particularly in terms of applications." As with U.S. fulfillment of almost outlandish production goals during the war, the success of the Manhattan District project, the U.S. performance against the Marshall Plan, and U.S. attainment of the Moon goal "within the decade," the Kremlin and the Soviet people generally know that the U.S. will succeed in meeting any objectives it sets for itself with High Frontier.

APPENDIX H: DISCUSSION OF IMPLEMENTING MEASURES

Implementation of High Frontier strategy depends on acquiring new civil and military space and active defense systems. In great measure, the time and cost to acquire these will determine whether the proposed strategy can effectively contribute to solving some of America's national security and economic problems. Acquisition of these new systems under existing methods, policies, and procedures will take too long and be too costly. It now takes 10 to 13 years to acquire a new system, which is unacceptably long for the proposed High Frontier systems. It is also costly. Literally billions can be saved by reducing acquisition time.

Aerospace systems such as Atlas and Polaris were selected, developed, and acquired in the 1950s in four to six years. When selected for development these systems had more technological unknowns than do the proposed High Frontier systems.

Of the numerous studies made during the past 10 years on the Defense system-acquisition problem, two have had a major impact and have resulted in subsequent actions. They are: (1) the 1977 Defense Science Board Summer Study Report, chaired by Dr. Richard DeLauer, now Undersecretary of Defense for Research and Engineering and (2) the April 30, 1981 Memorandum on Improving the Acquisition Process by Deputy Secretary of Defense Frank Carlucci. Both studies are in general agreement as to why the acquisition cycle has more than doubled in length since the 1950s.

The causes of today's overly long acquisition cycles have been identified; there is little or no disagreement on these. Corrective actions in all cases have been recommended and efforts are under way to implement these. Furthermore, precedent exists for shorter acquisition cycles since these continue to be successfully pursued in the case of some intelligence systems and commercial programs. Unfortunately, notwithstanding the recognized need by DOD to drastically reduce systems acquisition time, so far layerization, overregulation, and bureaucratic resistance continue to limit the progress in reducing this time to levels which are needed for High Frontier systems.

In addition, this major new Presidential initiative, with its national security goals, calls for a highly visible new management organization. Without such an organization, the effort would quickly acquire the image of just another service or departmental proposal to build a few new NASA and Defense space vehicles or weapons.

Therefore, if the earliest possible operational capabilities are to be achieved and a Presidential initiative identity maintained, the first generation of High Frontier systems should be selected and acquired under special organizational and procedural arrangements. Reducing the acquisition time will also reduce overall program costs. However, these special arrangements need only be such as to ensure that the systems acquisition process can benefit from all of the previously identified, agreed, and recommended measures in the DeLauer and Carlucci studies, while providing for the national stature of the program and the interdepartmental and possibly international nature of the effort.

Three basic alternative approaches were examined for acquiring the first generation of systems called for by the concept. These were:

- To establish a new, separate, centralized organization to manage all aspects of the acquisition of all the High Frontier systems by incorporating the initiatives to streamline the system acquisition process as outlined in Chapter VII.
- To establish a special task force to select the desired systems and then assign the acquisition task, in toto, to the department having primary interest in each system.
- To establish a centralized, interdepartmental organization which would not only select the specific systems to be acquired for the President but also would follow through on their acquisition by the responsible departments by retaining certain continuing policy, funding, public information, and other responsibilities.

In brief, the choice was to do the systems selection development and procurement tasks and to defend and justify the programs under a new organization in toto, under existing departments, or under a combination of the foregoing.

DISCUSSION OF THE ALTERNATIVES

Acquisition under a separate, new, centralized organization would ensure maximum exploitation of all shortcuts to include quick, decisive selection; effective funding; outstanding top management; certain priority; immediate decisions when needed; and freedom from in-house competition, turf-guarding opposition, and military service or other special interest influences. It would also best accommodate the interdepartmental and, if so decided, international nature of High Frontier and its special status as a bold Presidential initiative.

The government's human and physical resources and the administrative and research support required to acquire successfully any major high technology systems now lie mostly within the Department of Defense or NASA. Duplicating these or transfering them to new, centralized organizations would be time consuming and costly. It undoubtedly would be strongly resisted within the bureaucracy and would suffer from disinterest or refusal to serve on the part of many of the experts needed whose careers are tied to their departments or military services. Major legislative problems would also be created in connection with the authorizing, funding, and manning of new, separate, organizations whose mission would clearly conflict with that of both DOD and NASA.

On balance it would seem that any savings in time or enhancement of the effort's image, resulting from centralizing full acquisition management in new organizations, would be more than offset by the personnel, administrative, legislative, and morale problems generated and the delays in trying to resolve these.

Under the second alternative considered, selecting the desired systems and then delegating all High Frontier implementing activities to the departments, the President would approve the concept, appoint the proposed systems selection task force to pick the first generation systems to be acquired, and then disband the task force and direct the existing departments to fund, defend, and acquire the designated systems as a matter of priority.

This is the simplest solution if (1) the departments can be depended on to acquire the new systems expeditiously without continuing centralized supervision and (2) some provision could be made for the high visibility and interdepartmental (and possibly international) nature of the effort.

Unfortunately, as evidenced from DOD acquisition statistics and related studies, the departments have been unable to implement all the measures they themselves recognize as required for rapid and cost effective system acquisition. The size of the departments, the many organizational layers (especially in DOD), the compromising that is inherent in the joint service structure, inevitable turf-guarding of programs that will suffer from High Frontier priority funding, the requirement to operate within existing policies, procedures, and regulations, and interservice rivalries would quickly reduce the High Frontier systems to "just another program competing for funds and management skills," regardless of both Presidential and Secretary level initiatives or directives. Under normal departmental management High Frontier would quickly lose its character as a "bold, new, Reagan initiative."

In view of the shortcomings inherent in the first two acquisition-management alternatives considered, the third alternative—assignment to the departments of the acquisition function, with selected responsibilities for certain aspects of the High Frontier program delegated to new, centralized organizations—appears to be the best means of meeting all the vital requirements of the proposed initiative.

OVERALL DIRECTION

The urgency of fielding the first generation systems in order to cope with the Soviet threat and the need to overcome the delays inherent in departmental acquisition efforts call for separate, top level, overall direction of the effort. Concurrently, the fact that the departments have the majority of the U.S. facilities, expertise, and personnel resources required for large new system acquisition calls for their being assigned this task. The solution that appears best to achieve the desired goals and meet legitimate objections would be one in which:

- Overall policy direction, specific system selection, acquisition and allocation of funding, granting of priorities, rapid decisionmaking where necessary above departmental level, and public and congressional relations would be assigned to centralized organizations.
- Specific system-acquisition management, however, would be assigned to the interested departments.

Under this solution the centralized organizations would consist of: (1) a National Space Council under the chairmanship of the Vice President, (2) a Systems Selection Task Force (SSTF) reporting directly to the President, and (3) a Space Consolidated Program Office (SCPO) also reporting to the President but through the Chairman of Council.

These organizations would be established and defined by Executive Order. NASA would be designated executive agent for housekeeping for the new, centralized offices. Consideration should also be given to having the deputy administrator of NASA as head of the SCPO.

Specific systems for initial procurement would then be selected and recommended to the President for urgent acquisition by the SSTF.

Executive Orders, prepared by the SCPO for the President, would direct the responsible departments and agencies to undertake the acquisition of those systems selected as a matter of priority.

Specific authorizations, exceptions to procurement laws or regulations, priorities, and appropriate Presidential guidance, etc. would accompany these Executive Orders, as necessary, to ensure that specific system-acquisition management personnel in implementing departments have the freedom and support to vary from current acquisition procedures when this could save time or money.

The National Space Council would oversee the High Frontier initiative and the activities of the SCPO. In addition, it would provide a source of quick and final decisions, when and where needed, during the acquisition process.

THE SYSTEMS SELECTION PROCESS

The DOD studies of acquisition processes referred to earlier have indicated that great savings in overall acquisition time can be obtained by minimizing the time it now takes to agree upon the operational requirements and the specific systems to be developed to meet these decisions. These are the "front end" decisions for any new program.

The period from the identification of a requirement until full system development is approved (historically called the conceptual phase, phase O, or the definition phase) has steadily increased over the past two decades. Today it frequently exceeds six years. This can be attributed to a lack of decisiveness from top leadership as to what is wanted and the substitution of excessive studies, debates, analyses, and reviews. The current approach to defining the requirement and what will be built to meet these would introduce unacceptable delays in the availability of space and defense capabilities called for by High Frontier.

As a result of a review of the overall state of the art and of many industry proposals, it is clear that the desired systems can now be expeditiously selected, built, and deployed, and given adequate priority, funding, and special management arrangements. The President can make these decisions on the advice of the SSTF in lieu of waiting for industry or departmental formal and voluminous proposals. When and if made, the President's decisions can and should be accompanied by a commitment to procurement from the outset. These Presidential front end decisions could save up to six years in acquiring High Frontier systems.

In the 1950s a situation existed similar to what we face today. At that time, the Soviet threat called for the development of U.S. intercontinental missiles, but there were competing DOD requirements and industry proposals. As a result, DOD, and specifically the U.S. Air Force, lacked the motivation and/or technological confidence to direct full scale development and procurement of the first generation ICBM system concepts. President Eisenhower then appointed the Von Neumann Committee, consisting of recognized scientific and defense authorities, to select the specific ICBMs to be built at that time. On the basis of their advice he then directed the development and procurement of three systems: the Atlas, Titan, and Thor.

Our recommendation that President Reagan establish a High Frontier SSTF to review the various technical, service, and industry proposals for first generation space systems to implement this concept was based largely on this historically successful approach to quickly initiating a new national high technology program.

Such a SSTF would be a one time, ad hoc, review and selection board that would report directly to the President (see Figures 14 and 28). When the operational requirements are established by the approval of the High Frontier proposal the job of the SSTF would be solely to select and recommend to the President the specific systems that the departments should be immediately directed to acquire in order to implement these. The SSTF should be made up of leading U.S. aerospace scientists, industrialists, and management experts with such other technical expert members as the President might desire. We estimate a membership of between 9 and 12 individuals, with staff support provided by the SCPO discussed below.

The role of the SSTF would not be to debate the Assured Survival concept, or nature of this initiative, but *only* to recommend systems that can be quickly obtained to achieve the operational capabilities called for. The SSTF should be requested to render its recommendations within four months of activation and would be disbanded thereafter.

THE NATIONAL SPACE COUNCIL

In order to ensure coordinated support and policy direction for the High Frontier initiative at the highest level and to allay any concerns within the Executive Branch that the centralized organizations would usurp their responsibilities, it is proposed to establish a National Space Council somewhat similar to the one established under President Eisenhower in 1958.

The Council would oversee the implementation of the High Frontier effort on behalf of the President. It would provide broad policy guidance to the departments involved and direct the activities of the SCPO. The chairman of the Council would be the Vice President. He would act as the chief executive officer of the High Frontier Program as well as its principal liaison with the leaders of Congress in seeking their support and the funding for the program.

The Eisenhower-era council performed a useful role when the Soviet Sputnik event provided both a threat and an opportunity. Today we again face both a threat from and opportunity in space, hence, a similar Presidential Council, in conjunction with the High Frontier initiative, is deemed appropriate.

THE SPACE CONSOLIDATED PROGRAM OFFICE (SCPO)

A separate and independent SCPO should be established to: (1) ensure high program visibility, (2) ensure staff and management capability for centralized functions and as a source for rapid decisionmaking at the highest levels, and (3) provide for the interdepartmental, and possibly international, nature of the High Frontier program.

Specific responsibilities of the director of the SCPO should include the following:

- Provide a highly visible focal point for implementation of the Reagan High Frontier initiative and act as secretariat to the Council and SSTF.
- Draft Executive Orders and guidelines for departmental acquisition of selected High Frontier systems.
- Seek from Congress, justify, obtain, defend, and allocate the funding for the selected systems.
- Seek and obtain Presidential, Office of Management and Budget (OMB), or Congressional waivers of laws or regulations as necessary for departments to expedite acquisition.
- Work with the OMB and the Congress to reprogram funds among the various High Frontier system program offices where and when desirable.
- Resolve interdepartmental issues or refer these to the National Space Council, as appropriate.
- Provide or seek from the Council any decisions requested by the departments to expedite the program.
- Serve on any systems acquisition and review committees established by the departments to oversee High Frontier programs.
- Install and operate a management information system to monitor High Frontier progress and provide the President and the Council with frequent status reports.
- Coordinate with the Department of State any international negotiations relating to High Frontier.

DEPARTMENTAL ROLES

The departments will be assigned the responsibility to develop and acquire those specific High Frontier systems as directed by the President. This will be done by Executive Orders directing acquisition of each system selected.

The departments should be requested to set up special High Frontier offices at appropriate levels to expedite decisionmaking. Direct communications as needed with the centralized organizations should also be authorized to achieve these objectives. The goal should be to organize the effort so as to eliminate from the chain of command and review any organizations or staff levels not specifically assigned High Frontier acquisition responsibilities and/or having the authority to make final decisions thereon.

The Executive Orders to acquire specific systems might usefully specify the following roles for the responsible departments:

- Undertake the full scale development, test, evaluation, procurement, and operation of designated systems.
- Accomplish the foregoing in accordance with the priorities and policies assigned by the President and Council and by setting up special system program offices (SPOs) to staff these tasks with outstanding people.
- Adopt special management procedures and communication channels and delegate authority as necessary to permit High Frontier systems to benefit from all recommended acquisition actions in recent studies.
- Cooperate with the SCPO to establish direct communication channels and to implement a centralized management information system.
- Ensure that the funds appropriated for High Frontier acquisition are "fenced" and disbursed per the National Space Council's policies and directives.
- Support the National Space Council and the SCPO in their assigned missions.

The above department guidelines are illustrative. Actual guidelines would be discussed with department secretaries and reviewed by the Council before inclusion in Executive Orders. These orders would be prepared at the time of acquisition responsibility assignment for each system since they will differ between departments and systems. The goals to be met are set forth in detail in the "Prerequisites for Rapid and Efficient Acquisition," set forth in Tab A to this Appendix.

APPENDIX H—TAB A: PREREQUISITES FOR RAPID AND EFFICIENT ACQUISITION

KEY ACTIONS

1. The President assigns the highest priority to a space consolidated program.
2. The President announces the urgent need for and savings from establishing a space consolidated program and periodically explains to the nation how an extraordinary effort in space will provide our margin of safety.
3. The President announces the appointment of a Systems Selection Task Force (SSTF) with a finite life, like the "Von Neumann Committee" of the 1950s, composed of recognized scientific, defense, and acquisition authorities. Its purpose is to review candidate space systems and recommend to him, within a few months, the best systems to pursue.
4. The President announces the appointment of a National Space Council with a finite life span, consisting of high level membership from the White House, National Security Council, NASA, DOD, and others. Its purpose is to oversee the progress of acquiring selected systems and to resolve Congressional and interdepartmental issues.
5. The President announces the establishment of a Space Consolidated Program Office (SCPO) to provide centralized integration and direction for the acquisition of the selected High Frontier systems.
6. The President instructs all government agencies involved in this initiative to use realistic inflation factors in cost planning.

ORGANIZATION

1. The President issues Executive Orders and Presidential Directives establishing the SSTF, the National Space Council, and the SCPO.
2. The President selects membership for the Council and SSTF and issues an Executive Directive to NASA identifying NASA as the executive agent for housekeeping and logistic support of the new, centralized organization.
3. The Council's chairman, the Vice President, selects a director for the SCPO with national stature and experience who should report directly to him. The program office will also serve as secretariat to the Space Council and to the SSTF in addition to its other assigned functions.
4. Departments assigned acquisition responsibilities establish system program offices (SPOs) to acquire the selected systems and assign High Frontier expediters at departmental and systems-command levels.
5. The director determines the type of systems, engineering, and technical assistance support required by their SPOs traditionally provided by, but not limited to, Federal Contract Research Center (FCRCs). These should be under the command and control of the SPO managers.
6. Departments direct the government representative at major contractor plants to be responsive to the appropriate system manager on High Frontier programs.
7. Liaison personnel from operational users and from major contractors are assigned and colocated at the SPOs.
8. The High Frontier Systems Acquisition and Review Committee is established to include the director of the SCPO as chairman, with

members from responsible department head offices, system commands, or NASA centers, and the interested systems manager is established. This is to furnish the DOD Defense System Acquisition and Review Committee (DSARC) role and provide quick decisions when needed.

PERSONNEL AND CONTINUITY

1. The departments select and assign the best available talent in their departments to the SPOs.
2. All consolidated program and SPO personnel are assigned for the duration of their respective systems (approximately four to seven years) and in manning positions above their current grade to provide for in-place promotion.

DECISIONMAKING PROCESS

1. The President and Council seek an agreement with Congressional leadership to rely on the full reporting to and monitoring by the SCPO and Council in lieu of the prevalent micromanagement by Congressional staff and subcommittees that causes program delays.
2. The departments are authorized to place program objectives memoranda and budget cycles on a multiyear basis to be forwarded by the Council's chief executive officer to Congress for disposition.
3. Authorization is granted to the departments to delete determination and findings from the budget process. The SCPO director works out a streamlined spending authorization mechanism to be approved by the Council chairman and Congress as necessary.
4. Congress is persuaded to provide multiyear funding which will be "fenced" at department level, as in intelligence programs, and provides needed reprogramming authority for the SCPO director to include additional funds for the systems office to handle uncertainties and unforeseen technical problems.
5. Changes to overall system performance not resolved by the Acquisition Review Committee are referred to the Council chairman who acts on advice of the SCPO program director. As a ground rule all levels consider performance requirements sacred or unalterable if significant time and costs can be saved while providing "sufficient" system performance to obtain an Assured Survival capability.
6. First national priority is assigned for critical, exotic, limited availability and long lead time materials required for development, test, and production of selected systems.
7. The Council and departments delegate maximum authority and flexibility, as proposed in all recent acquisition studies, grant direct communication with any and all government decision levels, and free system managers from having to go to, or through, any official who cannot provide decisions.
8. Approval of the SSTF recommendations constitutes Milestone Zero. Those systems then approved for acquisition by the President go directly into DSARC II, full scale development phase, by Executive Order with full intent to deploy these systems.
9. Concurrency is the rule consistent with risk throughout the acquisition cycle, especially between full scale development and production, and in initiating and developing ground, logistic, training, and personnel support; command, control, and communications subsystems; and system facilities. Production is continued prior to completing the operational test and evaluation phase, and joint user/developer/contractor tests are conducted with full evaluation.
10. Departments are granted authority to waive existing DOD and OMB directives and regulations for acquisition management; re-

quests for proposals; statements of work; and system, subsystem, and end-item requirements specifications. Use directors' and managers' judgments in keeping with the spirit of the directives and regulations rather than applying a literal or stringent interpretation.

11. SCPO director's and system managers' discretion is applied for fair, quick, and flexible source selection procedures, with approval as needed from the Council chairman. The same flexibility applies to the form of contracts used for each system and subsystem, where all types are considered and carefully selected to best match the needs of each particular item and phase of the acquisition cycle.

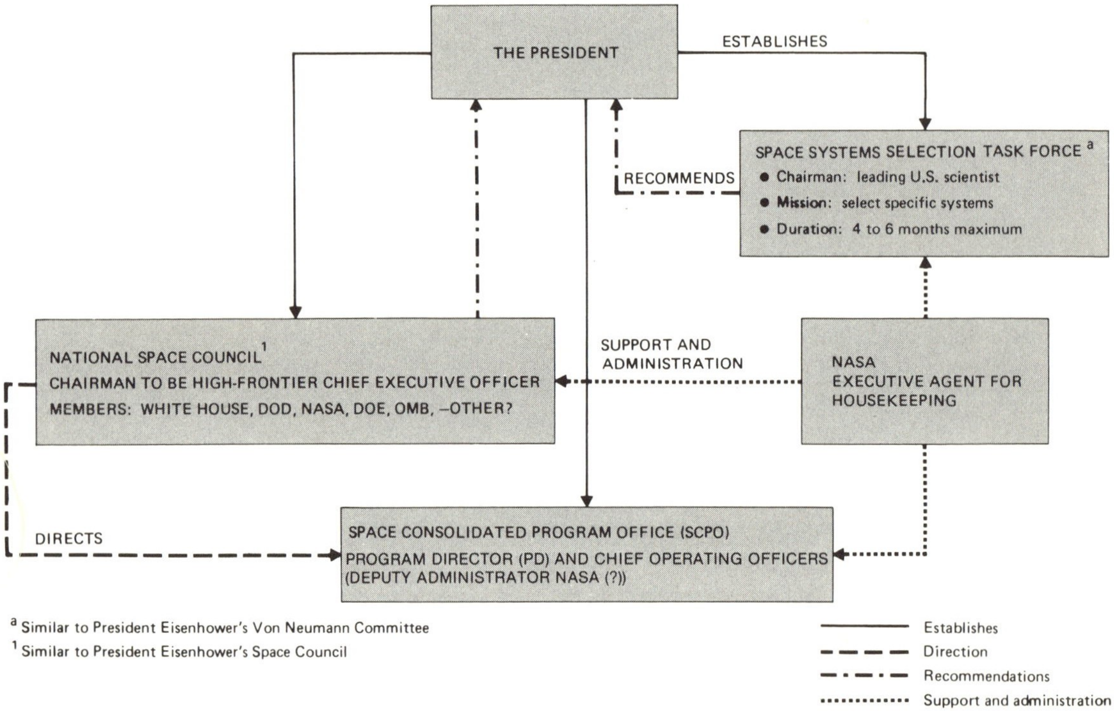

Figure 28. Proposed High Frontier Acquisition Organization—Top Echelon